Uwe Schäfer

From Sperner's Lemma to Differential Equations in Banach Spaces

An Introduction to Fixed Point Theorems and their Applications

From Sperner's Lemma to Differential Equations in Banach Spaces

An Introduction to Fixed Point Theorems and their Applications

by
Uwe Schäfer

KIT Scientific Publishing

Priv.-Doz. Dr. Uwe Schäfer

Karlsruher Institut für Technologie (KIT)
Institut für Angewandte und Numerische Mathematik
Uwe.Schaefer@kit.edu

Impressum

KIT Scientific Publishing

Karlsruher Institut für Technologie (KIT)
KIT Scientific Publishing
Straße am Forum 2
D-76131 Karlsruhe

KIT Scientific Publishing is a registered trademark of Karlsruhe Institute of Technology. Reprint using the book cover is not allowed.

www.ksp.kit.edu

Print on Demand 2014

ISBN 978-3-7315-0260-9
DOI 10.5445/KSP/1000042944

to Gabriele Schäfer

Preface

In this book we present several classical fixed point theorems and show how they have been applied. We further indicate where they are still applied to this very day.

We start with Brouwer's fixed point theorem. A well-known application by J. Nash, proving the existence of an equilibrium in any two-person game is presented in some detail.

Next, we present the Poincaré-Miranda theorem (by applying the Brouwer fixed point theorem) and then apply this theorem to prove the correctness of an algorithm used to solve nonlinear complementary questions, which arise, for example, in free boundary problems.

In the next chapter we deal with so-called verification methods. The idea is to use Brouwer's fixed point theorem as well as the Poincaré-Miranda theorem to verify, by use of a computer, that within a neighborhood of an approximate solution there really exists a solution of the problem $f(x) = 0$, where $f : \mathbb{R}^n \to \mathbb{R}^n$.

Chapter 7 deals with Banach's Contraction Mapping Principle. The classical uniqueness result for Cauchy's problem is derived from this famous principle.

Next, we present Schauder's fixed point theorem which can be viewed as a natural generalization of Brouwer's fixed point theorem to an infinite-dimensional setting. The classical Peano existence theorem is presented. We also deal with verification methods in relation to the existence of a solution of a semilinear elliptic partial differential equation in a neighborhood of an approximate solution. These methods were developed in the 1990s and exploit interval arithmetic on a computer. They still find use today in scientific research helping prove existence via the computer.

We close with Tarski's fixed point theorem. Though often overlooked, it is very elegant and we show how it helps prove the existence of a solution to a parabolic partial differential equation in a Banach space.

The book is based on a series of lectures on fixed point theorems and their applications that were given at the KIT (Karlsruhe Institute of Technology, formerly known as Univerität Fridericiana Karlsruhe (TH)). The idea for this series of lectures arose from the fact that fixed point theorems are used in the verification methods mentioned above. It seems that this application is not known, even to experts in fixed point theory.

Most authors of books about fixed point theory are functional analysts, algebraic topologists or mathematical economists. They do not consider numerical methods and so the problem of roundoff errors that occur during calculations goes unmentioned. As a consequence, no mention is made of verified results implemented on a computer using interval arithmetic via either the Brouwer or the Schauder fixed point theorems.

Rather than presenting the many beautiful proofs of Brouwer's fixed point theorem, we opt to expose many nice applications in some detail. It is amazing that fixed point theorems can be used in such different fields of mathematical endeavor and we wish to share how this is so with other mathematicians.

We introduce Sperner's lemma via a board game for children and derive Brouwer's fixed point theorem from it. Once that is done the following chapters can be treated more or less independently.

I thank Prof. Dr. Joseph Diestel from the Kent State University, Ohio for his interest, advice, and extremely valuable comments. His assiduous proofreading turned up many infelicities of grammar, logic, and style.

My thanks are also directed to Regine Tobias, to Brigitte Maier, and to Nadine Klumpp, of KIT Scientific Publishing, for supporting this project.

Karlsruhe, October 2014 Uwe Schäfer

Contents

1 Sperner's Lemma

Suppose two children are playing a board game. The board is a big triangle that is subdivided into smaller triangles as depicted in Figure 1.1. At the starting point of the game a red chip lies at the left vertex, a green chip lies at the right vertex, and a blue chip is lying at the vertex at the top.

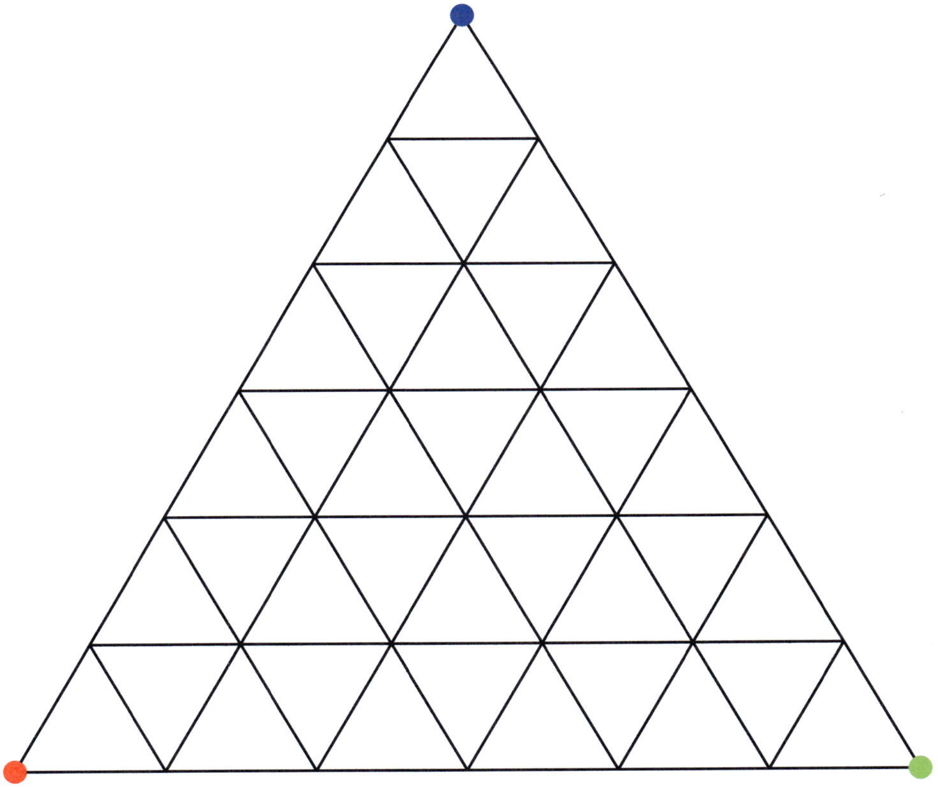

Figure 1.1: **Starting point of the game**

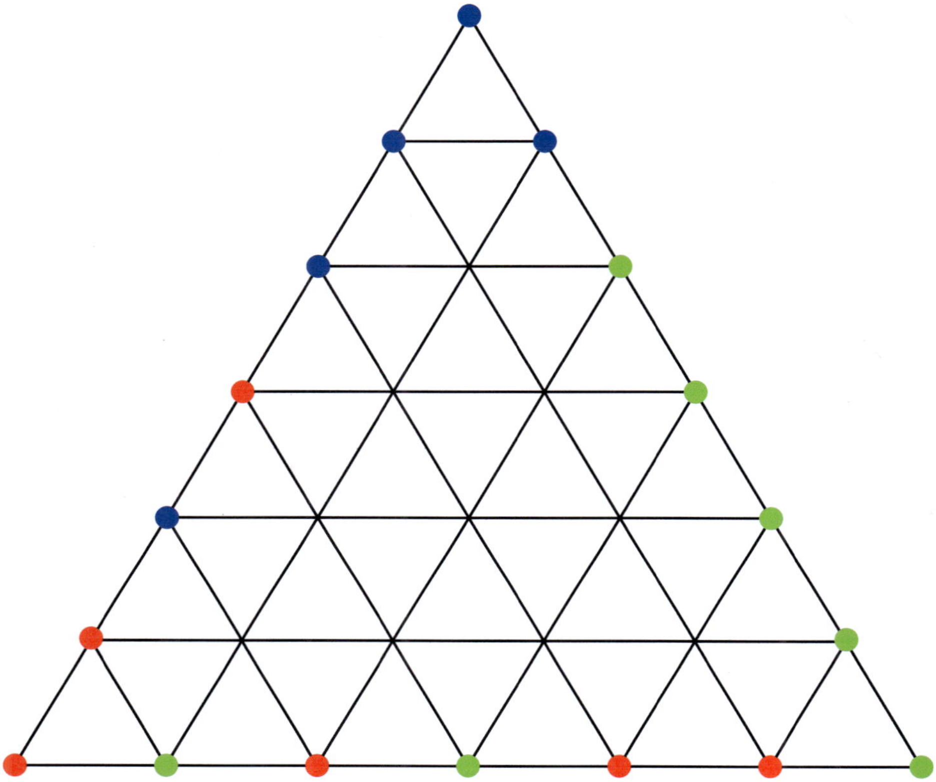

Figure 1.2: **The rule for the border edges**

The rules of the game are as follows. Both children are placing alternatively chips at the remaining vertices. (It is assumed that enough blue, red, and green chips are available). The children can place their chips at vertices whereever they want, but at the border edges of the big triangle there is a certain rule.

At the vertices lying on the left border edge of the big triangle it is only allowed to place red or blue chips. See Figure 1.2. On the other hand, at the vertices lying on the right border edge of the big triangle it is only allowed to place green or blue chips. Finally, at the vertices lying on the border edge at the bottom of the big triangle it is only allowed to place red or green chips.

Now, the aim of the game is to avoid a complete colored small triangle as long as possible, where a complete colored small triangle is a small triangle with a

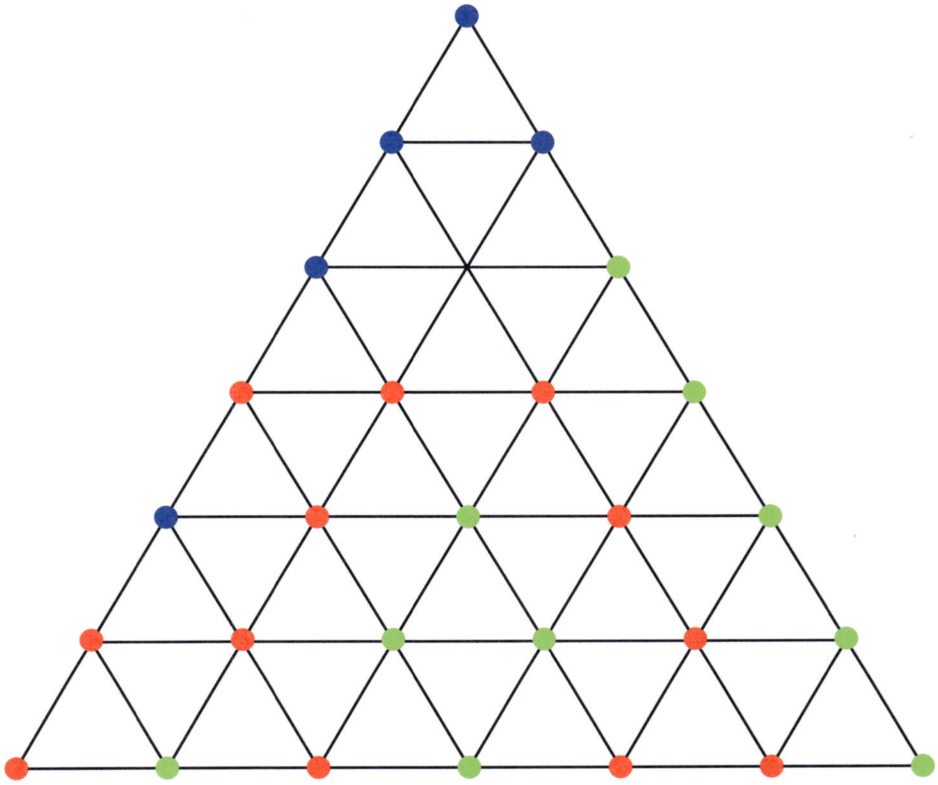

Figure 1.3: **Only one vertex is left**

blue, a red, and a green chip. In other words, the child who places a chip so that a complete colored small triangle arises is the loser.

Let us consider Figure 1.3. Up to this point, both children have played very well, because no complete colored small triangle has arisen, yet. However in this situation the child who has to place the last chip will lose. For example, if he (or she) places a green chip at the last vertex, then a complete colored small triangle arises as depicted in Figure 1.4. Also a red or a blue chip will lead to a complete colored small triangle. The reader can check this on his own.

The question arises: is there always a loser? In other words, will there arise a complete colored small triangle during the game? This is true, indeed. In fact, it is a consequence of Sperner's lemma in the two-dimensional case.

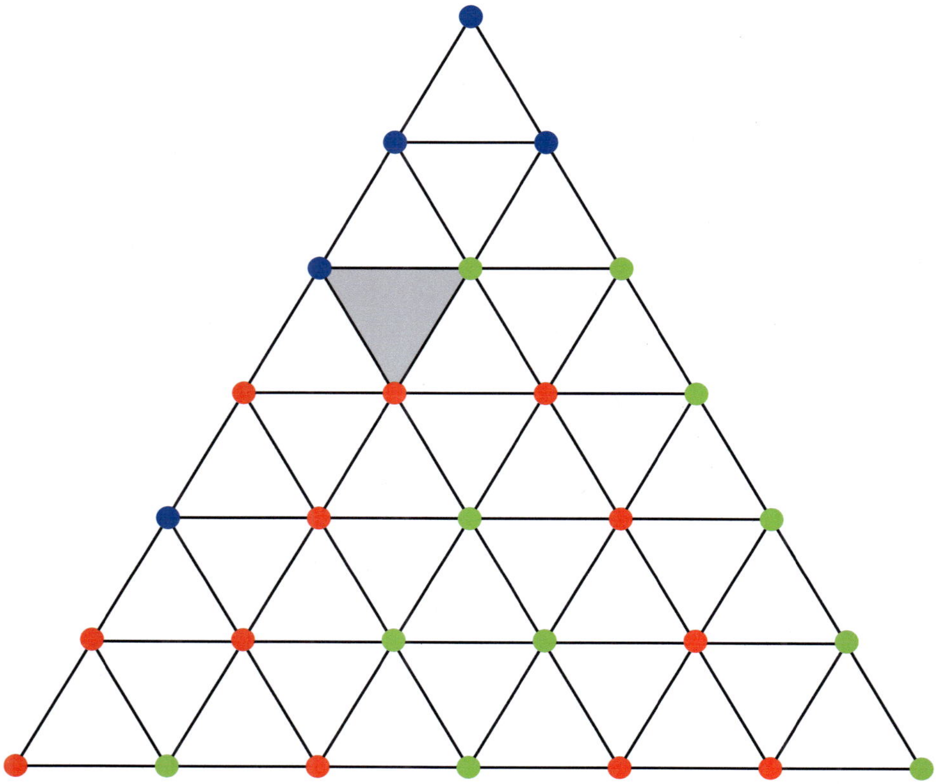

Figure 1.4: **A complete colored small triangle has arisen**

How can we prove it? Let us consider Figure 1.4. Assume, we start a path from the bottom of the big triangle, where it is only allowed to cross an edge of a small triangle for the case that the two vertices of this edge are red and green. We will show, that such a path either will end again at the bottom or it will reach a complete colored small triangle. See Figure 1.5.

Since the vertices lying on the left border edge of the big triangle are blue and red, and since the vertices lying on the right border edge of the big triangle are blue and green, such a path can leave the big triangle neither on the left border edge nor on the right border edge.

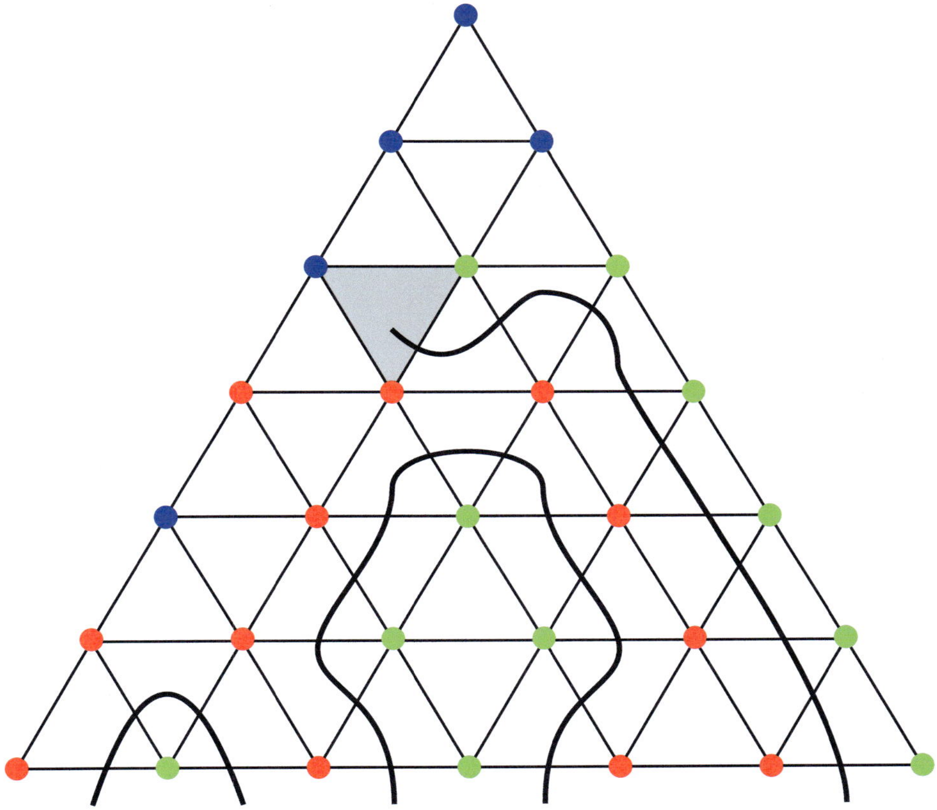

Figure 1.5: **The path that reaches a complete colored small triangle**

Furthermore, such a path cannot cross three edges of the same small triangle, because two colors of the three vertices must be the same, if the path crosses twice the same small triangle. As a consequence, such a path cannot have a loop. So, such a path either will end again at the bottom or it will reach a complete colored small triangle. See again Figure 1.5.

The remaining thing that we have to show is that there exists at least one path that will not end again at the bottom of the big triangle.

To verify this, consider the border edge at the bottom of the big triangle.

Going from left to right, since at the beginning of the game at the left vertex there was a red chip and since at the right vertex there was a green chip, there must be a small edge with a red chip at the left and a green chip at the right. If afterwards there is also a small edge with a green chip at the left and a red chip at the right, then there must be a corresponding small edge with a red chip at the left and a green chip at the right.

Otherwise the border edge at the bottom will not end at a green vertex.

As a consequence, at the bottom of the big triangle the number of small edges consisting of a red chip and a green chip must be odd. (Concerning this number it does not matter if the red chip is on the left and the green chip is on the right or vice versa.)

See Figure 1.5. At the bottom of the big triangle the number of small edges consisting of a red and a green chip is 5.

So, at least one path that starts at an edge with a red chip and a green chip at the bottom of the big triangle will not come back to the bottom. This path must end in a complete colored small triangle.

This means that the board game indeed ends with a loser. Actually, Sperner's lemma says more: Assume the children's game is not finished when the first complete colored small triangle occurs but when the last chip is set.

A variant of the children's game could be that this child who created more complete colored small triangles than the other one has lost the game.

Also in this variant of the game there is always a loser because - and this is Sperner's lemma in the two-dimensional case - the number of complete colored small triangles is odd. See Figure 1.6.

To prove this we just continue the proof for the fact that there is at least one complete colored small triangle. Up to this point we have proved the following two statements:

(S1) The number of edges with a red chip and a green chip at the bottom of the big triangle is odd.

(S2) Each path from the bottom of the big triangle, where it is only allowed to cross an edge of a small triangle for the case that the two chips are green and red, either will end again at the bottom of the big triangle or it will reach a complete colored small triangle.

With these two statements we can conclude that the number of complete colored small triangles that were reached by a path started from the bottom of the big triangle is odd.

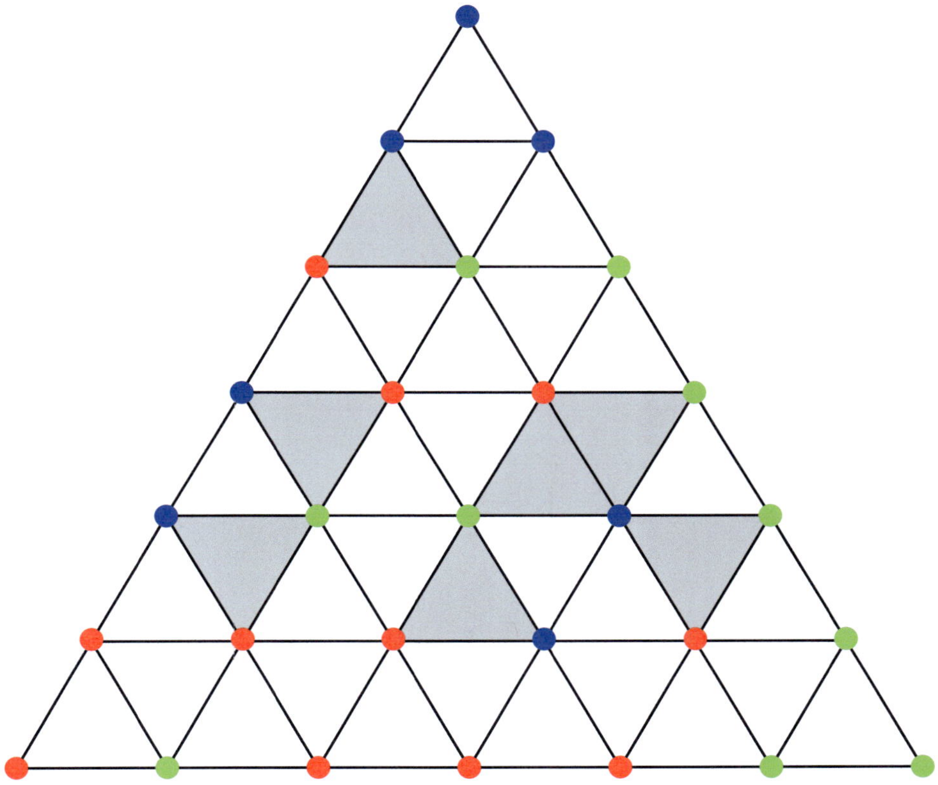

Figure 1.6: **Seven complete colored small triangles**

Now, if there is another complete colored small triangle that was *not* reached by a path that started from the bottom of the big triangle, then a path that starts from this complete colored small triangle (where it is only allowed to cross an edge of a small triangle for the case that the two chips are green and red) will end in a further complete colored small triangle that was also not reached by a path that started from the bottom of the big triangle. See Figure 1.7.

This means that complete colored small triangles that are not reached by a path started from the bottom of the big triangle arise pairwise, and so we have proved a third statement:

(S3) The number of complete colored small triangles that are not reached by a path started from the bottom of the big triangle is even.

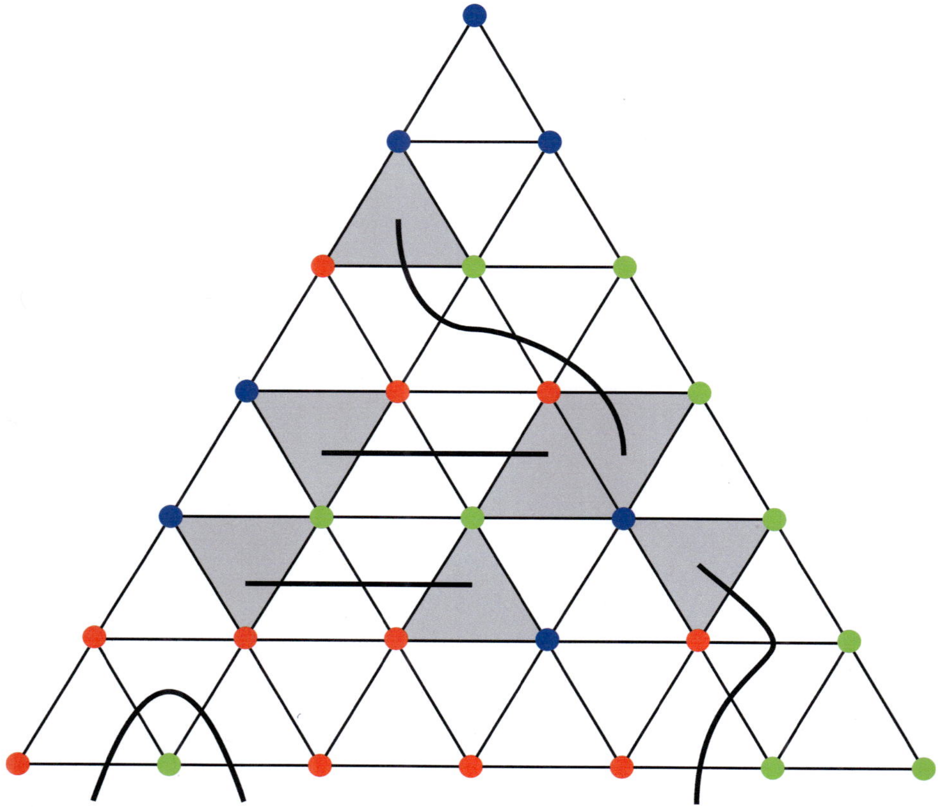

Figure 1.7: **Even further complete colored triangles**

Together with the two statements (S1) and (S2) it follows that the number of complete colored small triangles is odd.

1.1 Simplexes and Barycentric Coordinates

In geometry, a simplex (plural simplexes or simplices) is a generalization of the notion of a triangle or tetrahedron to arbitrary dimensions. Specifically, a k-simplex is a k-dimensional polytope which is the convex hull of its k + 1 vertices. More formally, in 2 dimensions a simplex is a triangle; in 1 dimension a simplex is a line segment; in N dimensions a simplex is the set of all points P

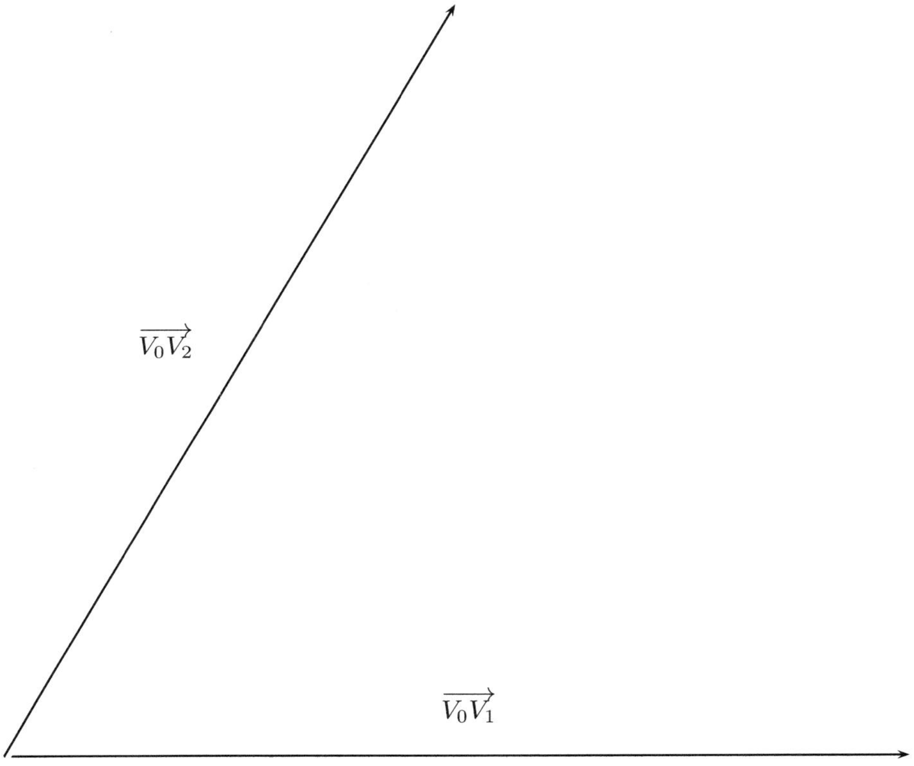

Figure 1.8: **The two vectors spanning the simplex**

with

$$\overrightarrow{OP} = x_0 \cdot \overrightarrow{OV_0} + x_1 \cdot \overrightarrow{OV_1} + ... + x_N \cdot \overrightarrow{OV_N}$$

so that

$$x_0 \geq 0,\ x_1 \geq 0, ..., x_N \geq 0 \quad \text{and} \quad x_0 + x_1 + ... + x_N = 1.$$

Here, $V_0, ..., V_N$ are the vertices of the simplex, and $\overrightarrow{OV_0}, \overrightarrow{OV_1}, ..., \overrightarrow{OV_N}$ are the associated position vectors. The simplex is called nondegenerate, if the N vectors

$$\overrightarrow{V_0V_1},\ \overrightarrow{V_0V_2},\ ...,\ \overrightarrow{V_0V_N}$$

are linearly independent. The numbers $x_0,\ x_1, ..., x_N$ are the barycentric coordinates of the point P.

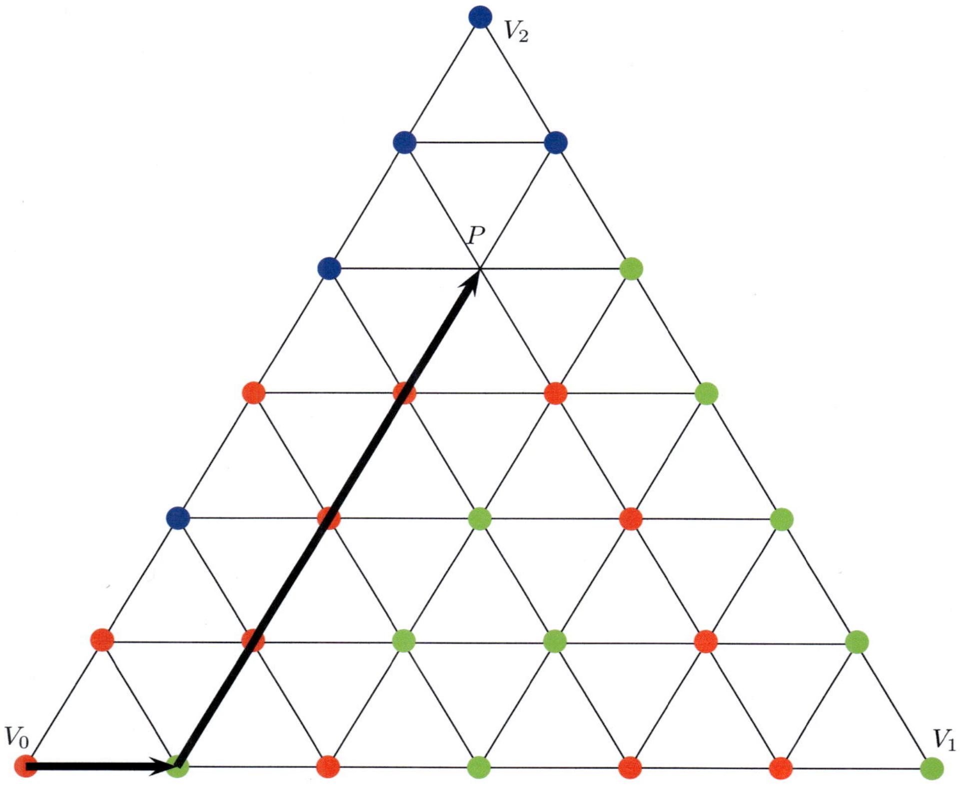

Figure 1.9: **Figure 1.3 combined with the vector addition (1.1)**

Example 1.1. Consider Figure 1.1. It is a simplex in 2 dimensions with the 6th subdivision. Let the red vertex be V_0, let the green vertex be V_1, and let the blue vertex be V_2. Since the vectors $\overrightarrow{V_0V_1}$ and $\overrightarrow{V_0V_2}$ are obviously linearly independent, the simplex is nondegenerate. See also Figure 1.8.

Each point within the simplex can be reached via a linear combination of $\overrightarrow{V_0V_1}$ and $\overrightarrow{V_0V_2}$. For example, let in Figure 1.3 the point P be the vertex without a color. Then, (see also Figure 1.9)

$$\overrightarrow{V_0P} = \frac{1}{6} \cdot \overrightarrow{V_0V_1} + \frac{4}{6} \cdot \overrightarrow{V_0V_2}. \tag{1.1}$$

Via

$$\overrightarrow{V_0V_1} = \overrightarrow{OV_1} - \overrightarrow{OV_0}, \quad \overrightarrow{V_0V_2} = \overrightarrow{OV_2} - \overrightarrow{OV_0}, \quad \text{and} \quad \overrightarrow{V_0P} = \overrightarrow{OP} - \overrightarrow{OV_0}$$

we get from (1.1)

$$\overrightarrow{OP} = \frac{1}{6} \cdot \overrightarrow{OV_0} + \frac{1}{6} \cdot \overrightarrow{OV_1} + \frac{4}{6} \cdot \overrightarrow{OV_2}.$$

This means, that the vertex P has the barycentric coordinates

$$\left(\frac{1}{6}, \frac{1}{6}, \frac{4}{6}\right).$$

□

In general, for $n = 2, 3, ...$ we form the nth barycentric subdivision of the simplex as follows. The new vertices are the points with barycentric coordinates

$$\left(\frac{k_0}{n}, \frac{k_1}{n}, ..., \frac{k_N}{n}\right),$$

where the k_i are integers satisfying

$$\text{all } k_i \geq 0 \quad \text{and} \quad \sum_{i=0}^{N} k_i = n.$$

The original simplex is the whole body; the subsimplexes are its cells. In general, in the nth subdivision of a simplex in N dimensions, the number of cells tends to infinity as $n \to \infty$; but the diameter of each cell tends to zero. If Δ is the diameter of the body, then the diameter of each cell is $\frac{\Delta}{n}$.

1.2 The Proof of Sperner's Lemma

Sperner's lemma concerning simplexes reads as follows.

Theorem 1.1. *Form the nth barycentric subdivision of a nondegenerate simplex in N dimensions. Label each vertex with an index $i \in \{0, 1, ..., N\}$ that satisfies the following restriction on the boundary.*

> *Each vertex from the vertices V_p, V_q, ..., V_s, that define a boundary element, is labeled with an index $i \in \{p, q, ..., s\}$.*

Then, the number of cells in the subdivision, that have vertices with the complete set of labels: $0, 1, ..., N$, is odd.

Proof: The proof is by induction, where the case $N = 1$ is trivial and the case $N = 2$ was done via the board game for children.

For the general proof we use some notation. If the vertices of a cell have the labels $m_0, m_1, ..., m_k$, we say the cell has type $(m_0, m_1, ..., m_k)$. Here permutations do not matter, but multiplicities do matter.

A cell in N dimensions has boundary elements of dimensions $0, 1, ..., N-1$. Those of dimension 0 are vertices; those of dimensions $N-1$ we will call faces. We will call the whole cell an element of dimension N.

By $F(a, b, ..., q)$ we mean the number of elements in the body of type $(a, b, ..., q)$. Let us look at the faces labeled $(0, 1, ..., N-1)$. Two such faces occur in each cell of type

$$(0, 1, ..., N-1, m) \quad \text{if } m \in \{0, 1, ..., N-1\}.$$

On the other hand, one face labeled $(0, 1, ..., N-1)$ occurs in each cell of type

$$(0, 1, ..., N-1, N).$$

Now look at the sum

$$2 \cdot \sum_{m=0}^{N-1} F(0, 1, ..., N-1, m) + F(0, 1, ..., N-1, N).$$

This sum, abbreviated by S, counts every interior face $(0, 1, ..., N-1)$ twice, since every interior face is shared by two cells; but the sum counts every face $(0, 1, ..., N-1)$ on the boundary of the body only once, because each face on the boundary belongs to only one cell. Therefore, the sum S satisfies

$$S = 2 \cdot F_i(0, 1, ..., N-1) + F_b(0, 1, ..., N-1).$$

Sperner's lemma applies with dimension $N-1$, and so $F_b(0, 1, ..., N-1)$ is odd by induction. Therefore, S is odd, which implies that $F(0, 1, ..., N-1, N)$ is odd. □

Remark 1.1. Emanuel Sperner (9 December 1905 - 31 January 1980) was a German mathematician. The idea to present and to prove his lemma via a board game for children is from the DVD: *Das Spernersche Lemma* by W. Dröge, H. Göttlich, and F. Wille published 1983 by the IWF Göttingen. □

2 Brouwer's Fixed Point Theorem

The Brouwer fixed point theorem was published by him in 1912. Three years before, Brouwer could only prove his theorem for $n = 3$, and it was Hadamard who in 1910 gave the first proof for arbitrary n.[1]

Brouwer's proof differs from that of Hadamard and meanwhile, there are many other proofs. See, for example, any of the books J. Franklin (32), J. Appell and M. Väth (9) or H. Heuser (46). Brouwer's fixed point theorem is a topological result, and most proofs are topological. Here, we decided to present the proof of Knaster-Kuratowski-Mazurkiewicz given in 1929 using the lemma of Sperner.

This chapter is organized as follows. First, we state the fixed point theorem and we present some examples to illustrate it. Then, we give some examples to show that no assumption can be dropped. In following sections we give the proof of the theorem in two steps. First, we give the proof for simplexes and then, based on this, we give the proof for convex sets.

Formally, the theorem reads as follows.

Theorem 2.1. (Brouwer's fixed point theorem) *Let K be a nonempty closed bounded convex subset of $\mathbb{R}^n$ and suppose f is a continuous function from K to itself. Then, there exists some $x^* \in K$ so that $f(x^*) = x^*$.*

Example 2.1. Consider Figure 2.1. There you can see the graph of a function $f : [a, b] \to [a, b]$. Being a self-mapping means, that the graph must stay within the square $[a, b] \times [a, b]$. Being continuous means, that the graph has to cross the function $y = x$. So, Brouwer's fixed point theorem is more or less obvious for the case $n = 1$. □

Of course, there can be more than one fixed point. See Figure 2.2.

It is not that easy, to verify Brouwer's fixed point theorem in higher dimensions. For example, consider a cup of coffee. Imagine, you pour carefully the coffee from one cup to anoter cup in one go, where both cups have equal size and equal form. Then, Brouwer's fixed point theorem says that at least one particle of the coffee will be at the same position. It is not an easy task to verify this.

[1] See (55).

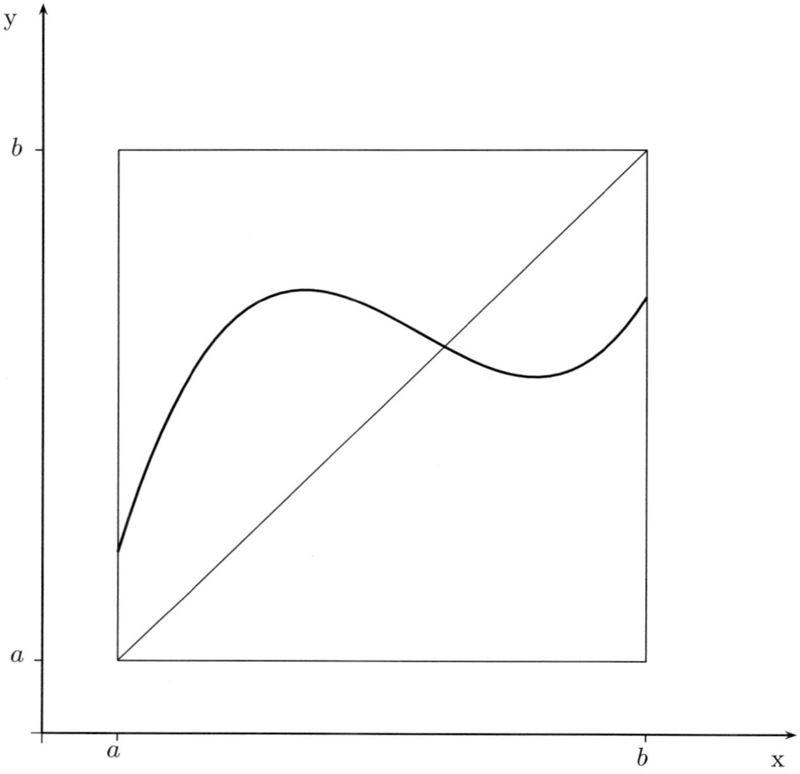

Figure 2.1: **One fixed point**

At least it is not as easy as using Figure 2.1 in order to verify Brouwer's fixed point theorem for $n = 1$.

Before we give a proof of Brouwer's fixed point theorem, we give some examples to show that Brouwer's fixed point theorem is no longer true if some assumptions are dropped.

Before, let us recall the definition of a norm in $\mathbb{R}^n$. A function $\|\cdot\| : \mathbb{R}^n \to \mathbb{R}$ is called a norm in $\mathbb{R}^n$ if it fulfills the following three conditions:

$$\begin{aligned} \|x\| &\geq 0 \text{ and } \|x\| = 0 \Leftrightarrow x = 0, \\ \|\lambda \cdot x\| &= |\lambda| \cdot \|x\|, \\ \|x + y\| &\leq \|x\| + \|y\|. \end{aligned}$$

Here, $x, y \in \mathbb{R}^n$ and $\lambda \in \mathbb{R}$. Note, that we use 0 for the number zero as well as for the zero vector.

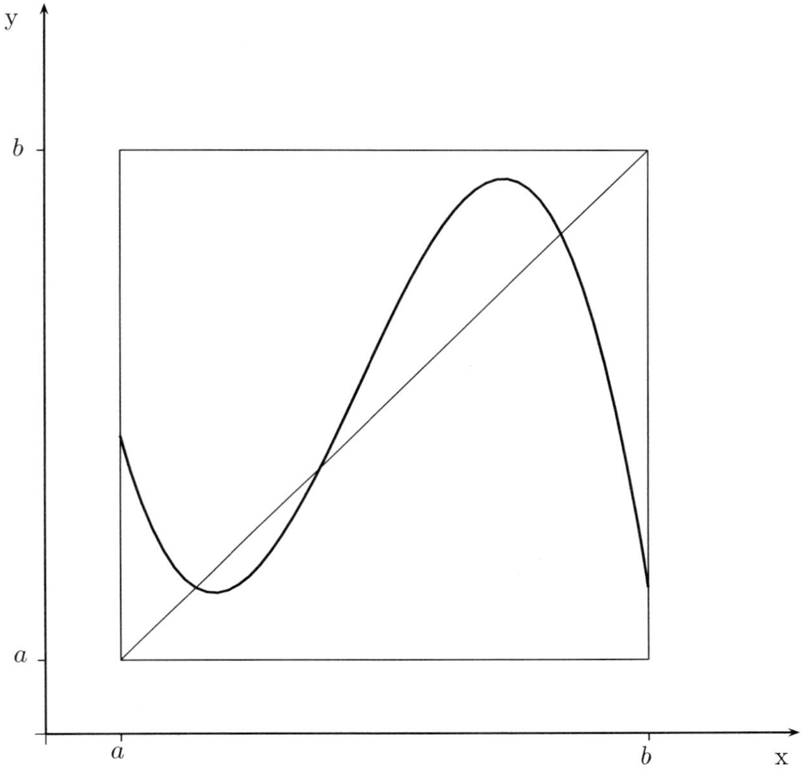

Figure 2.2: **More fixed points**

Perhaps the most popular norm in $\mathbb{R}^n$ is the Euclidean norm which is defined as follows:

$$\|x\|_e := \sqrt{\sum_{j=1}^{n} x_j^2} \quad \text{with} \quad x = \begin{pmatrix} x_1 \\ \vdots \\ x_n \end{pmatrix} \in \mathbb{R}^n.$$

Theorem 2.2. *All norms in $\mathbb{R}^n$ are equivalent; i.e., for any two different norms $\|\cdot\|_1$ and $\|\cdot\|_2$ in $\mathbb{R}^n$ there exist two positive numbers α, β so that*

$$\alpha \cdot \|x\|_1 \leq \|x\|_2 \leq \beta \cdot \|x\|_1 \quad \textit{for all} \quad x \in \mathbb{R}^n.$$

Proof: It suffices to show that any norm $\|\cdot\|$ in $\mathbb{R}^n$ is equivalent to the Euclidean norm. So, let $\|\cdot\|$ be an arbitrary norm in $\mathbb{R}^n$ and let e_j denote the vector in

$\mathbb{R}^n$ with a 1 in the jth entry and 0's elsewhere. Then using

$$\max_{j=1}^{n} |x_j| = \max_{j=1}^{n} \sqrt{x_j^2} \leq \sqrt{\sum_{j=1}^{n} x_j^2}$$

we get

$$\|x\| = \|\sum_{j=1}^{n} x_j \cdot e_j\| \leq \sum_{j=1}^{n} |x_j| \cdot \|e_j\| \leq \beta \cdot \|x\|_e$$

with

$$\beta = \sum_{j=1}^{n} \|e_j\|.$$

Assume there is no $\alpha > 0$ so that

$$\alpha \cdot \|x\|_e \leq \|x\|$$

is valid for all $x \in \mathbb{R}^n$. Then, for each $k > 0$ there is some $x^{(k)} \in \mathbb{R}^n$ with

$$\frac{1}{k} \cdot \|x^{(k)}\|_e > \|x^{(k)}\|.$$

Dividing by $\|x^{(k)}\|_e$ and setting

$$y^{(k)} := \frac{1}{\|x^{(k)}\|_e} \cdot x^{(k)}$$

we get a sequence $\{y^{(k)}\}_{k=1}^{\infty}$ satisfying

$$\frac{1}{k} \cdot \|y^{(k)}\|_e > \|y^{(k)}\| \quad \text{and} \quad \|y^{(k)}\|_e = 1 \quad \text{for all } k \in \mathbb{N}. \tag{2.1}$$

Due to the theorem of Bolzano-Weierstraß we can conclude that there is a convergent subsequence $\{y^{(k_l)}\}_{l=1}^{\infty}$ concerning $\|\cdot\|_e$; i.e., there exists some $y^* \in \mathbb{R}^n$ with

$$\lim_{l \to \infty} \|y^{(k_l)} - y^*\|_e = 0.$$

Since

$$\|y^{(k_l)} - y^*\| \leq \beta \cdot \|y^{(k_l)} - y^*\|_e$$

this subsequence is also convergent to y^* concerning the norm $\|\cdot\|$. On the one hand, (2.1) implies $\|y^*\|_e = 1$, whence $y^* \neq 0$. On the other hand, this contradicts to (2.1) because of $\frac{1}{k_l} > \|y^{(k_l)}\|$. □

Example 2.2. (concerning boundedness) A set K in $\mathbb{R}^n$ is bounded if there is a constant M so that

$$\|x\| \leq M \quad \text{for all } x \in K$$

for some norm $\|\cdot\|$ in $\mathbb{R}^n$. Due to Theorem 2.2 it is irrelevant concerning which norm this is valid.

Let $K = [0, \infty)$ and $f : K \to K$ defined by $f(x) = x + 1$. Then f has no fixed point at all. □

Example 2.3. (concerning closedness) A set K in $\mathbb{R}^n$ is closed if every convergent sequence in K has its limit in K.

Let $K = (0, 1)$ and $f : K \to K$ defined by $f(x) = x^2$. Then due to

$$f(x) = x \quad \Leftrightarrow \quad x \cdot (x - 1) = 0$$

f has two fixed points $x^* = 0$ and $x^{**} = 1$, but neither belongs to K. □

Example 2.4. (concerning convexity) A set K in $\mathbb{R}^n$ is convex if

$$x, y \in K \quad \Rightarrow \quad \lambda \cdot x + (1 - \lambda) \cdot y \in K$$

for all $\lambda \in [0, 1]$ and for all $x, y \in K$. Let

$$K = \{x = \begin{pmatrix} x_1 \\ x_2 \end{pmatrix} \in \mathbb{R}^2 : \frac{1}{2} \leq \sqrt{x_1^2 + x_2^2} \leq 1\}.$$

K is just a disc with a hole and therefore not convex:

$$x = \begin{pmatrix} -\frac{3}{4} \\ 0 \end{pmatrix}, y = \begin{pmatrix} \frac{3}{4} \\ 0 \end{pmatrix} \in K \quad \text{but} \quad K \not\ni \begin{pmatrix} 0 \\ 0 \end{pmatrix} = \frac{1}{2} \cdot x + (1 - \frac{1}{2}) \cdot y.$$

Let $f : K \to K$ be defined by rotating K by 45 degree to the left. Then the zero vector is the only fixed point of f. It does not belong to K. □

The next example shows that the phrase 'f is continuous' actually has to read 'f is continuous subject to some norm', since the theorem was wrong if the assumption would be changed to 'f is continuous subject to some metric'.

Example 2.5. (concerning continuity subject to some norm) A function $f : \mathbb{R}^n \to \mathbb{R}^n$ is continuous subject to some norm $\|\cdot\|$ if for any $x^* \in \mathbb{R}^n$

$$\lim_{k \to \infty} \|f(x^{(k)}) - f(x^*)\| = 0 \quad \text{if} \quad \lim_{k \to \infty} \|x^{(k)} - x^*\| = 0. \tag{2.2}$$

Due to Theorem 2.2 it is irrelevant which norm is used in (2.2).

Let $\mathbb{R}^n$ be equipped with the so-called discrete metric which is defined as follows.

$$d(x^*, \hat{x}) := \begin{cases} 1 & \text{if} \quad x^* \neq \hat{x}, \\ 0 & \text{if} \quad x^* = \hat{x}. \end{cases}$$

Continuity subject to this metric concerning any function f means, that from

$$\lim_{n\to\infty} d(x^{(n)}, x^*) = 0$$

it follows

$$\lim_{n\to\infty} d(f(x^{(n)}), f(x^*)) = 0.$$

However, according to this property any function f is continuous as we will see. Let

$$\lim_{n\to\infty} d(x^{(n)}, x^*) = 0.$$

This means, that there exists some $n_0 \in \mathbb{N}$ so that

$$x^{(n)} = x^* \qquad \text{for all } n \geq n_0.$$

As a consequence,

$$f(x^{(n)}) = f(x^*) \qquad \text{for all } n \geq n_0.$$

So,

$$\lim_{n\to\infty} d(f(x^{(n)}), f(x^*)) = 0.$$

Now, let $x_0,\ y_0 \in \mathbb{R}^n$, $x_0 \neq y_0$ and let

$$K := \{x \in \mathbb{R}^n : x = x_0 + \lambda \cdot (y_0 - x_0) \text{ for some } \lambda \in [0, 1]\}.$$

Then, K is nonempty, bounded, closed, convex, and the function $f : K \to K$ defined by

$$f(x) := \begin{cases} x_0 & \text{if} \quad x \neq x_0 \\ y_0 & \text{if} \quad x = x_0 \end{cases}$$

is continuous subject to the discrete metric, but there is no fixed point. □

2.1 The Proof for Simplexes

We assume that the reader is familiar with Section 1.1.

Lemma 2.1. *Let S be a simplex, and let $x, y \in S$ be given in barycentric coordinates. Then,*

$$x \geq y \quad \Rightarrow \quad x = y.$$

Proof: Assume, that $x \neq y$, although $x \geq y$. Then, there exists some i so that $x_i > y_i$. Hence, due to

$$1 = \sum x_i > \sum y_i = 1,$$

we get a contradiction. □

Theorem 2.3. (Brouwer's fixed point theorem for simplexes) *Let $S \subseteq \mathbb{R}^N$ be a nondegenerate simplex in N dimensions, and let $f : S \to S$ be continuous. Then, there exists some $x^* \in S$ so that $x^* = f(x^*)$.*

Proof: For each $x \in S$ we have $f(x) \in S$. Therefore,

$$\sum_{i=0}^{N} x_i = 1 = \sum_{i=0}^{N} f_i(x). \tag{2.3}$$

Assume, for all $x \in S$ it holds

$$f(x) \neq x.$$

Then, for each $x \in S$ there must be some i satisfying $x_i \neq f_i(x)$, and, due to (2.3), then there must be some j satisfying

$$x_j > f_j(x).$$

So, for each $x \in S$ we define

$$m(x) := \min\{j : 0 \leq j \leq N,\, x_j > f_j(x)\}.$$

Let n be a large positive integer. Forming the nth barycentric subdivision of S, we see that $m(x)$ has the following property. If x belongs to a boundary element S_F of S, then those coefficients of

$$x = x_0 \cdot \overrightarrow{OV_0} + x_1 \cdot \overrightarrow{OV_1} + ... + x_N \cdot \overrightarrow{OV_N}$$

that belongs to vertices not in S_F are zero. That means, if

$$V_p,\, V_q, ..., V_s \in S_F, \quad \text{and for the remaining vertices } V_i,\, V_i \notin S_F,$$

then

$$x = x_p \cdot \overrightarrow{OV_p} + x_q \cdot \overrightarrow{OV_q} + ... + x_s \cdot \overrightarrow{OV_s}.$$

Because $x_j > f_j(x) \geq 0$ it follows that

$$m(x) \in \{p, q, ..., s\} \quad \text{for all } x \in S_F.$$

Now, Sperner's lemma says, that there is some cell with $N+1$ vertices

$$X_0(n),\ X_1(n), ..., X_N(n)$$

satisfying

$$\begin{aligned} m &= 0 \quad \text{at the vertex } X_0(n), \\ m &= 1 \quad \text{at the vertex } X_1(n), \\ &\vdots \\ m &= N \quad \text{at the vertex } X_N(n). \end{aligned}$$

Then, by definition of $m(x)$ we have

$$\left.\begin{aligned} x_0 &> f_0(x) \quad \text{for } x = \overrightarrow{OX_0(n)}, \\ x_1 &> f_1(x) \quad \text{for } x = \overrightarrow{OX_1(n)}, \\ &\vdots \\ x_N &> f_N(x) \quad \text{for } x = \overrightarrow{OX_N(n)}. \end{aligned}\right\} \tag{2.4}$$

We can do this for any n that's large. Since S is bounded and closed, there is a subsequence n_l so that there is $X^* \in S$ and

$$\lim_{l\to\infty} \overrightarrow{OX_0(n_l)} = \overrightarrow{OX^*}.$$

Furthermore, all the vertices of the cell are close to each other if n is large. So,

$$\lim_{l\to\infty} \overrightarrow{OX_j(n_l)} = \overrightarrow{OX^*} \quad \text{for all } j \in \{0, 1, ..., N\}.$$

Now, the continuity of f implies

$$\lim_{l\to\infty} f(\overrightarrow{OX_j(n_l)}) = f(\overrightarrow{OX^*}) \quad \text{for all } j \in \{0, 1, ..., N\}.$$

So, we get in (2.4)

$$\begin{aligned} x_0^* &\geq f_0(x^*), \\ x_1^* &\geq f_1(x^*), \\ &\vdots \\ x_N^* &\geq f_N(x^*). \end{aligned}$$

Finally, Lemma 2.1 leads to $x^* = f(x^*)$ with $x^* = \overrightarrow{OX^*}$. □

2.2 The Proof for Convex Sets

The step from a simplex to an arbitrary convex set is done using the following definition.

Definition 2.1. *Two sets A, B are called topologically equivalent if there is a continuous function $\Psi : A \to B$, with a continuous inverse $\Psi^{-1} : B \to A$.*

We present an example in the following theorem.

Theorem 2.4. *Let $K \subseteq \mathbb{R}^N$ be a closed bounded convex set with nonempty interior. Then, K is topologically equivalent to a closed ball in $\mathbb{R}^N$.*

Proof: Let B be an interior point of K and let u be any unit vector in $\mathbb{R}^N$. Then, K contains a boundary point $\overrightarrow{OB} + \lambda \cdot u$ with $\lambda > 0$. For each u there is only one such boundary point. Why? Suppose there were two with $\lambda_1 < \lambda_2$. Remember that K contains some small ball Q centered at B. But then the point P_1 with $\overrightarrow{OP_1} = \overrightarrow{OB} + \lambda_1 \cdot u$ lies interior to the convex hull of the ball Q and the point P_2 with $\overrightarrow{OP_2} = \overrightarrow{OB} + \lambda_2 \cdot u$. Therefore, P_1 is interior to K, which is impossible for a boundary point.

Thus, for each u in $\mathbb{R}^N$, there is a unique

$$\lambda = \lambda(u) > 0$$

so that P with $\overrightarrow{OP} = \overrightarrow{OB} + \lambda \cdot u$ lies on the boundary of K. The function $\lambda(u)$ is called the radial function.

We assert that the radial function is continuous as a function of the unit vector u. If $u^{(0)}$ were a point of discontinuity, then the boundary of K would contain points P_k with

$$\overrightarrow{OP_k} = \overrightarrow{OB} + \lambda_k \cdot u^{(k)}, \quad k = 1, 2, 3, ...$$

with $u^{(k)} \to u^{(0)}$ but with positive λ_k not converging to $\lambda(u^{(0)})$. Since λ_1, $\lambda_2, ...$ is a bounded sequence, there is a subsequence with a limit $\lambda^* \neq \lambda(u^{(0)})$. But then $\overrightarrow{OB} + \lambda^* u^{(0)}$ must lie on the boundary of K, since a limit of boundary points must be a boundary point, too. Now we would have two boundary points P_* and P_0 with

$$\overrightarrow{OP_*} = \overrightarrow{OB} + \lambda^* \cdot u^{(0)} \quad \text{and} \quad \overrightarrow{OP_0} = \overrightarrow{OB} + \lambda(u^{(0)}) \cdot u^{(0)}.$$

But this is impossible, since the boundary point in the direction $u^{(0)}$ is unique.

Now, we define the function $\Psi : K \to \{y \in \mathbb{R}^N : \|y\|_e \leq 1\}$ as follows:

$$\Psi(C) := \frac{1}{\lambda\left(\frac{\overrightarrow{BC}}{\|\overrightarrow{BC}\|_e}\right)} \cdot \overrightarrow{BC}, \quad \text{if} \quad C \neq B$$

and

$$\Psi(B) := 0.$$

On the other hand let $y \in \mathbb{R}^N$ with $\|y\|_e \leq 1$. Then,

$$\Psi^{-1}(y) := \overrightarrow{OB} + \lambda(\frac{y}{\|y\|_e}) \cdot y, \quad \text{if} \quad y \neq 0$$

and

$$\Psi^{-1}(0) := B.$$

By definition of the radial function $\lambda(\cdot)$ we have

$$\|\overrightarrow{BC}\|_e \leq \lambda\left(\frac{\overrightarrow{BC}}{\|\overrightarrow{BC}\|_e}\right) \quad \text{if} \quad C \neq B.$$

Therefore, if $C \neq B$

$$\|\Psi(C)\|_e = \frac{\|\overrightarrow{BC}\|_e}{\lambda\left(\frac{\overrightarrow{BC}}{\|\overrightarrow{BC}\|_e}\right)} \leq 1.$$

On the other hand,

$$\Psi^{-1}(\Psi(C)) = \overrightarrow{OB} + \lambda\left(\frac{\Psi(C)}{\|\Psi(C)\|_e}\right) \cdot \Psi(C)$$

$$= \overrightarrow{OB} + \lambda\left(\frac{\frac{\overrightarrow{BC}}{\lambda\left(\frac{\overrightarrow{BC}}{\|\overrightarrow{BC}\|_e}\right)}}{\left\|\frac{\overrightarrow{BC}}{\lambda\left(\frac{\overrightarrow{BC}}{\|\overrightarrow{BC}\|_e}\right)}\right\|_e}\right) \cdot \frac{\overrightarrow{BC}}{\lambda\left(\frac{\overrightarrow{BC}}{\|\overrightarrow{BC}\|_e}\right)} = \overrightarrow{OB} + \overrightarrow{BC} = \overrightarrow{OC}$$

if $C \neq B$. Since we have already shown that the radial function is continuous, the only thing that remains to be shown is

$$\lim_{k\to\infty} \|\Psi(C^{(k)})\|_e = 0, \quad \text{if} \lim_{k\to\infty} C^{(k)} = B$$

and

$$\lim_{k\to\infty} \Psi^{-1}(y^{(k)}) = \overrightarrow{OB}, \quad \text{if} \lim_{k\to\infty} y^{(k)} = 0.$$

However, due to

$$\lambda\left(\frac{\overrightarrow{BC^{(k)}}}{\|\overrightarrow{BC^{(k)}}\|_e}\right) \geq dist(B, boundary\, of\, K) > 0$$

and

$$\lambda\left(\frac{y^{(k)}}{\|y^{(k)}\|_e}\right) \leq diameter\, of\, K$$

this is obvious. □

Now, we are able to prove Brouwer's fixed point theorem. For completeness, we state it once more.

Theorem 2.5. (Brouwer's fixed point theorem) *Let $K \neq \emptyset$, $K \subseteq \mathbb{R}^n$ be bounded, convex, and closed. Furthermore, let $f : K \to K$ be continuous. Then, there exists some $x^* \in K$ so that $f(x^*) = x^*$.*

Proof: Case 1. We assume that K has dimension n. Then, by Theorem 2.4 K is topologically equivalent to a closed ball in $\mathbb{R}^n$; i.e., there exists a continuous function

$$\Phi : K \to \{y \in \mathbb{R}^n : \|y\|_e \leq 1\}$$

with a continuous inverse Φ^{-1}. Again, by Theorem 2.4 the set $\{y \in \mathbb{R}^n : \|y\|_e \leq 1\}$ is topologically equivalent to any nondegenerate n-simplex S with $S \subseteq \mathbb{R}^n$. Therefore, there exists a continuous function

$$\varphi : \{y \in \mathbb{R}^n : \|y\|_e \leq 1\} \to S$$

with a continuous inverse φ^{-1}. Setting

$$\Psi := \varphi \circ \Phi$$

we have that K is topologically equivalent to S with $\Psi : K \to S$. Now, we define

$$g := \Psi \circ f \circ \Psi^{-1}. \tag{2.5}$$

Then, $g : S \to S$ is continuous. By Brouwer's fixed point theorem for simplexes (see Theorem 2.3) g has a fixed point $z^* \in S$; i.e., $z^* = g(z^*)$. Via (2.5) we get

$$z^* = \Psi(f(\Psi^{-1}(z^*))).$$

Hence,

$$\Psi^{-1}(z^*) = f(\Psi^{-1}(z^*)).$$

This means, that $x^* := \Psi^{-1}(z^*)$ is a fixed point of f satisfying $x^* \in K$.

Case 2. We assume that K has dimension m with $m < n$. Then, there exists m orthogonal unit vectors $u^{(1)}, ..., u^{(m)}$ so that for all $x \in K$ there are uniquely defined numbers $y_1, ..., y_m$ so that

$$x = y_1 \cdot u^{(1)} + ... + y_m \cdot u^{(m)}.$$

With

$$U := \left(u^{(1)} \vdots ... \vdots u^{(m)} \right) \in \mathbb{R}^{n \times m} \quad \text{and} \quad y := \begin{pmatrix} y_1 \\ \vdots \\ y_m \end{pmatrix} \in \mathbb{R}^m$$

we have $x = U \cdot y$ and $U^T \cdot x = y$. Setting

$$\tilde{K} := \{y \in \mathbb{R}^m : y = U^T x \text{ with } x \in K\}$$

we have that $\tilde{K}$ is bounded, convex, and closed with dimension m. For any $y \in \tilde{K}$ we define

$$\tilde{f}(y) := U^T \cdot f(U \cdot y).$$

Then, $\tilde{f} : \tilde{K} \to \tilde{K}$ is continuous. Applying Case 1 with n replaced by m the function $\tilde{f}$ has a fixed point $y^* \in \tilde{K}$; i.e.,

$$y^* = \tilde{f}(y^*).$$

It follows

$$U \cdot y^* = f(U \cdot y^*).$$

This means, that $x^* := U \cdot y^*$ is a fixed point of f satisfying $x^* \in K$. $\square$

3 Nash's Equilibrium

Although the movie "A Beautiful Mind" from 2001 starts with the claim that 'mathematicians won World War II', the main focus of the movie is not on mathematicians but on the life of one particular mathematician, John Forbes Nash, Jr., whose contributions were post-World War II.

If you talk about this movie with friends and colleagues the question will arise about what John Nash did. Sooner or later you might have to explain the meaning of 'Nash equilibrium'. This chapter is aimed at helping you with such an explanation.

3.1 Two-Person Games

Concerning a two-person game two people (player 1 and player 2) are playing against each other. Both players have the ability to make their bets. We assume that player 1 can choose from m possibilities, whereas player 2 can choose from n (in general other) possibilities. To simplify, we assume that there are given two $m \times n$ matrices

$$A = \begin{pmatrix} a_{11} & a_{12} & \cdots & a_{1n} \\ a_{21} & a_{22} & \ddots & \vdots \\ \vdots & \ddots & \ddots & a_{m-1\,n} \\ a_{m1} & a_{m2} & \cdots & a_{mn} \end{pmatrix}, \quad B = \begin{pmatrix} b_{11} & b_{12} & \cdots & b_{1n} \\ b_{21} & b_{22} & \ddots & \vdots \\ \vdots & \ddots & \ddots & b_{m-1\,n} \\ b_{m1} & b_{m2} & \cdots & b_{mn} \end{pmatrix}$$

with the following meaning. For the case that player 1 places his bet on the ith possibility and player 2 places his bet on the jth possibility, then player 1's costs are defined by

$$a_{ij} = e_i^{\mathrm{T}} A e_j \,,$$

and player 2's costs are given by

$$b_{ij} = e_i^{\mathrm{T}} B e_j .$$

Sometimes both players are allowed to split up their bets into different possibilities. We give an example. Player 1 stakes half of his money on the first possibility and bets the other half of his money on the second possibility. Player 2 stakes a quarter of his money on the first possibility and he stakes the remaining money on the last possibility. Then, player 1's costs are defined by

$$(\frac{1}{2}\ \frac{1}{2}\ 0\cdots 0)A\begin{pmatrix}\frac{1}{4}\\ 0\\ \vdots\\ 0\\ \frac{3}{4}\end{pmatrix},$$

and player 2's costs are given by

$$(\frac{1}{2}\ \frac{1}{2}\ 0\cdots 0)B\begin{pmatrix}\frac{1}{4}\\ 0\\ \vdots\\ 0\\ \frac{3}{4}\end{pmatrix}.$$

This leads us to the definition of a mixed strategy.

Definition 3.1. *For $m, n \in \mathbb{N}$ we define*

$$\begin{aligned} S^m &:= \{x = (x_i) \in \mathbb{R}^m : x_i \geq 0,\ i = 1, ..., m,\ \sum_{i=1}^{m} x_i = 1\}, \\ S^n &:= \{y = (y_j) \in \mathbb{R}^n : y_j \geq 0,\ j = 1, ..., n,\ \sum_{j=1}^{n} y_j = 1\}. \end{aligned}$$

Any $x \in S^m$ is called mixed strategy concerning player 1, and any $y \in S^n$ is called mixed strategy concerning player 2. Especially, any $x \in S^m$ with $x_i = 1$ for some i is called pure strategy concerning player 1. Analogously, pure strategies concerning player 2 are defined.

Example 3.1. Suppose we have two software companies working together. One is called A from Argentina and the other one is called B from Brasil. One part of their cooperation is given by some data transmission from A to B and vice versa. Suppose in Argentina there are m different network providers, whereas in Brasil there are n different network providers. Furthermore, suppose, that the companies have agreed to the fact that company A will meet the costs in Argentina and company B will meet the costs in Brasil.

Since the prices of the network providers are given, there are two cost matrices $A, B \in \mathbb{R}^{m \times n}$ representing the costs as follows. If company A chooses the ith network provider for all data and if company B chooses the jth network provider for all data, then company A has to pay a_{ij} and company B has to pay b_{ij}.

Now the question arises how the data should be split up into the network providers so that the costs are as small as possible for both companies. □

3.2 Nash's Equilibrium

Concerning a two-person game the question naturally arises if there exists a pair of strategies so that the costs are as small as possible for both players. That is, one asks if $\hat{x} \in S^m$ and $\hat{y} \in S^n$ exist with

$$\hat{x}^{\mathrm{T}} A\hat{y} \leq x^{\mathrm{T}} Ay \quad \text{for all } x \in S^m,\ y \in S^n, \tag{3.1}$$

and

$$\hat{x}^{\mathrm{T}} B\hat{y} \leq x^{\mathrm{T}} By \quad \text{for all } x \in S^m,\ y \in S^n. \tag{3.2}$$

The following example will show that such a pair of strategies does not exist in general.

Example 3.2. Let

$$A = \begin{pmatrix} 5 & 0 \\ 10 & 1 \end{pmatrix} \quad \text{and} \quad B = \begin{pmatrix} 5 & 10 \\ 0 & 1 \end{pmatrix}.$$

Suppose, there are $\hat{x} \in S^2$ and $\hat{y} \in S^2$ with

$$\hat{x}^{\mathrm{T}} A\hat{y} \leq x^{\mathrm{T}} Ay \quad \text{for all } x \in S^2,\ y \in S^2 \tag{3.3}$$

and

$$\hat{x}^{\mathrm{T}} B\hat{y} \leq x^{\mathrm{T}} By \quad \text{for all } x \in S^2,\ y \in S^2. \tag{3.4}$$

We will show that from (3.3) it follows that

$$\hat{x} = \begin{pmatrix} 1 \\ 0 \end{pmatrix}, \quad \hat{y} = \begin{pmatrix} 0 \\ 1 \end{pmatrix}. \tag{3.5}$$

Firstly, for any $x \in S^2$ and for any $y \in S^2$ we have $x^{\mathrm{T}} Ay \geq 0$ because of $A \geq O$. Since the vectors from (3.5) fulfill $\hat{x}^{\mathrm{T}} A\hat{y} = 0$, it is sufficient to show

$$\hat{x}^{\mathrm{T}} A\hat{y} = 0 \quad \Rightarrow \quad \hat{x} = \begin{pmatrix} 1 \\ 0 \end{pmatrix}, \quad \hat{y} = \begin{pmatrix} 0 \\ 1 \end{pmatrix}.$$

So, let $\hat{x}^{\mathrm{T}} A\hat{y} = 0$. Then, from

$$0 = \hat{x}^{\mathrm{T}} A\hat{y} = 5\hat{x}_1\hat{y}_1 + 10\hat{x}_2\hat{y}_1 + \hat{x}_2\hat{y}_2$$

we get the following three conditions

$$5\hat{x}_1\hat{y}_1 = 0, \quad 10\hat{x}_2\hat{y}_1 = 0, \quad \hat{x}_2\hat{y}_2 = 0. \tag{3.6}$$

In particular, $\hat{x}_2\hat{y}_1 = 0$. Since $\hat{x}$, $\hat{y} \in S^2$, by (3.6) we get

$$\hat{y}_1 = 0 \Rightarrow \hat{y}_2 = 1 \Rightarrow \hat{x}_2 = 0$$

and

$$\hat{x}_2 = 0 \Rightarrow \hat{x}_1 = 1 \Rightarrow \hat{y}_1 = 0.$$

Therefore, we have $\hat{y}_1 = 0$ and $\hat{x}_2 = 0$, which implies (3.5). Secondly, if we substitute (3.5) in (3.4) we get a contradiction because

$$\hat{x}^{\mathrm{T}} B\hat{y} = 10 > 0 = (0\ 1)B \begin{pmatrix} 1 \\ 0 \end{pmatrix}.$$

□

Example 3.2 shows that, in general, it is possible that no optimal pair of strategies exist in a two-person game. So, it is no surprise that sometimes two companies have to abandon negotiations, since no agreement could be achieved, even if dealing was allowed.

A third person who wants to arbitrate can take advantage of Nash's equilibrium.

Definition 3.2. *Concerning a two-person game defined by* $A,\ B \in \mathbb{R}^{m\times n}$*, a pair of strategies* $(\hat{x}, \hat{y})$, $\hat{x} \in S^m$, $\hat{y} \in S^n$ *is called a Nash equilibrium if*

$$\begin{aligned} \hat{x}^T A\hat{y} &\leq x^T A\hat{y} \quad \text{for all } x \in S^m, \\ \hat{x}^T B\hat{y} &\leq \hat{x}^T By \quad \text{for all } y \in S^n \end{aligned}$$

is valid.

The knowledge of some Nash equilibrium can be exploited by some arbitrator as follows. The arbitrator is going to player 1 saying:

> "I know that player 2 is playing strategy $\hat{y}$. With this assumption it can be guaranteed mathematically that your costs are as small as possible if you play strategy $\hat{x}$."

Afterwards, the arbitrator is going to player 2 saying:

> "I know that player 1 is playing strategy $\hat{x}$. With this assumption it can be guaranteed mathematically that your costs are as small as possible if you play strategy $\hat{y}$."

Player 1 and player 2 are not talking to each other (in fact, they quarrel with each other). So, no player can say if he was contacted by the arbritrator first or not.

We want to emphasize that it is possible that there exists some pair of strategies $(\overline{x}, \overline{y})$ satisfying

$$\overline{x}^{\mathrm{T}} A\overline{y} < \hat{x}^{\mathrm{T}} A\hat{y} \quad \text{and} \quad \overline{x}^{\mathrm{T}} B\overline{y} < \hat{x}^{\mathrm{T}} B\hat{y}.$$

This means, that the main advantage of the knowledge of some Nash equilibrium is not taken by the players but is taken by the arbitrator!

Example 3.3. (Prisoners' dilemma) After a robbery two suspicious persons were arrested. The custodial judge, who wants these two persons to confess to this crime, is aware of some small piece of evidence. So, he tries to play one suspect off against the other one by offering a deal:

> If both suspects confess, then both suspects will be arrested for five years in prison. In case that exactly one suspect confesses, then the other suspect will be arrested for ten years in prison and the confessing suspect will be free. If both suspects deny, then both suspects will be arrested for one year in prison due to illicit possession of a firearm.

The custodial judge sends the two suspects in different cells, so they have to make their decisions independently from each other on how to testify. Now, the suspects are playing a two-person game.

In this game a Nash equilibrium is the situation that both suspects confess. The custodial judge exploits the knowledge of this Nash equilibrium by visiting the suspects consecutively by saying:

> "If I was you, I would confess, since your fellow has already confessed."

Neither suspect knows whether he was visited by the custodial judge first. The prisoners' dilemma is the fact that the custodial judge benefits from the knowledge of the Nash equilibrium but neither of the prisoners do; if both suspects claim innocence, both will fare better. □

3.3 Nash's Proof of Existence

In contrast to Example 3.2 where we have seen that some two-person games do not have an optimal solution for both players, any two-person game can be arbitrated by a Nash equilibrium because John Forbes Nash Jr. has shown that there exists a Nash equilibrium in any two-person game. We give the proof in the following theorem.

Theorem 3.1. *Given a two-person game defined by A, $B \in \mathbb{R}^{m \times n}$ there exists a Nash equilibrium.*

Proof: Let $A_{i\cdot}$ denote the ith row of A and let $B_{\cdot j}$ denote the jth column of B. Then, for $(x, y) \in S^m \times S^n$ and via

$$\begin{aligned} c_i(x,y) &:= \max\{x^{\mathrm{T}}Ay - A_{i\cdot}y, 0\}, \quad & i = 1, ..., m, \\ d_j(x,y) &:= \max\{x^{\mathrm{T}}By - x^{\mathrm{T}}B_{\cdot j}, 0\}, \quad & j = 1, ..., n, \end{aligned}$$

we define the continuous mapping $T : S^m \times S^n \to \mathbb{R}^m \times \mathbb{R}^n$ by $T(x,y) := (x', y')$ with

$$x'_i := \frac{x_i + c_i(x,y)}{1 + \sum_{k=1}^{m} c_k(x,y)}, \quad i = 1,...,m, \quad y'_j := \frac{y_j + d_j(x,y)}{1 + \sum_{k=1}^{n} d_k(x,y)}, \quad j = 1,...,n.$$

For $i = 1,...,m$, $c_i(x,y) \geq 0$ and so $\sum_{k=1}^{m} c_k(x,y) \geq 0$. From $x \in S^m$ then it follows that $x' \in S^m$. Analogously, one can show $y' \in S^n$. Therefore,

$$T(x,y) \in S^m \times S^n, \qquad \text{if } (x,y) \in S^m \times S^n. \tag{3.7}$$

Now, we show

$$T(x,y) = (x,y) \quad \Leftrightarrow \quad (x,y) \text{ is a Nash equilibrium.} \tag{3.8}$$

"$\Leftarrow$": Let $(\hat{x}, \hat{y})$ be a Nash equilibrium. Then, for all $i = 1,...,m$ it holds

$$\hat{x}^{\mathrm{T}} A \hat{y} \leq e_i^{\mathrm{T}} A \hat{y} = A_{i\cdot} \hat{y}.$$

So, we have $c_i(\hat{x}, \hat{y}) = 0$ for $i = 1,...,m$. By the same argument we get $d_j(\hat{x}, \hat{y}) = 0$ for $j = 1,...,n$. Therefore, we get $T(\hat{x}, \hat{y}) = (\hat{x}, \hat{y})$.
"$\Rightarrow$": We assume that $T(\hat{x}, \hat{y}) = (\hat{x}, \hat{y})$, although $(\hat{x}, \hat{y})$ is not a Nash equilibrium. Then, either there exists some $\overline{x} \in S^m$ with $\overline{x}^{\mathrm{T}} A \hat{y} < \hat{x}^{\mathrm{T}} A \hat{y}$ or there exists some $\overline{y} \in S^n$ with $\hat{x}^{\mathrm{T}} B \overline{y} < \hat{x}^{\mathrm{T}} B \hat{y}$. Since both cases lead to a contradiction in the same way, we assume without loss of generality, that for some $\overline{x} \in S^m$ it holds

$$\overline{x}^{\mathrm{T}} A \hat{y} < \hat{x}^{\mathrm{T}} A \hat{y}. \tag{3.9}$$

First, we will show that due to (3.9) it holds

$$A_{i\cdot} \hat{y} < \hat{x}^{\mathrm{T}} A \hat{y} \quad \text{for at least one } i \in \{1,...,m\}. \tag{3.10}$$

Assume, that (3.10) is not true. Then, it follows that $\min_{1 \leq i \leq m} A_{i\cdot} \hat{y} \geq \hat{x}^{\mathrm{T}} A \hat{y}$. Therefore,

$$\overline{x}^{\mathrm{T}} A \hat{y} \quad = \quad \sum_{i=1}^{m} \overline{x}_i A_{i\cdot} \hat{y} \geq \left(\min_{1 \leq i \leq m} A_{i\cdot} \hat{y} \right) \cdot \sum_{i=1}^{m} \overline{x}_i = \min_{1 \leq i \leq m} A_{i\cdot} \hat{y} \geq \hat{x}^{\mathrm{T}} A \hat{y}$$

contradicting (3.9). So, (3.10) is shown, and it follows $c_i(\hat{x}, \hat{y}) > 0$ for at least one $i \in \{1,...,m\}$. This implies

$$\sum_{k=1}^{m} c_k(\hat{x}, \hat{y}) > 0. \tag{3.11}$$

Second, we will show:

$$\text{There exists some } i \in \{1, ..., m\} \text{ with } \hat{x}_i > 0 \text{ and } c_i(\hat{x}, \hat{y}) = 0. \tag{3.12}$$

Suppose, (3.12) is not true. Then, for every $i \in \{1, ..., m\}$ it holds

$$\hat{x}_i = 0 \quad \text{or} \quad A_{i\cdot}\hat{y} < \hat{x}^{\mathrm{T}} A\hat{y}. \tag{3.13}$$

So, we get via

$$\hat{x}^{\mathrm{T}} A\hat{y} = \sum_{i=1}^{m} \hat{x}_i A_{i\cdot}\hat{y} < \sum_{i=1}^{m} \hat{x}_i(\hat{x}^{\mathrm{T}} A\hat{y}) = \hat{x}^{\mathrm{T}} A\hat{y}$$

a contradiction. Therefore, (3.12) is true. Concerning the index i from (3.12) we get togethter with (3.11)

$$x_i' = \frac{\hat{x}_i}{1 + \sum\limits_{k=1}^{m} c_k(\hat{x}, \hat{y})} < \hat{x}_i.$$

This means that $T(\hat{x}, \hat{y}) \neq (\hat{x}, \hat{y})$. As a consequence, (3.8) is shown.

We've shown the equivalence of the existence of a Nash equilibrium and the existence of a fixed point for the map T. Now the set $S^m \times S^n$ is a nonempty, closed bounded convex set and the function T is a continuous self-map on $S^m \times S^n$; so Brouwer's fixed point theorem tells us T has a fixed point and, with that, we also have a Nash equilibrium. □

Theorem 3.1 is a pure existence theorem. It presents no algorithm to calculate a Nash equilibrium. The calculation of a Nash equilibrium can be done via a so-called linear complementarity problem. This was done by C. E. Lemke and J. T. Howson in 1964.

Remark 3.1. If you watch the movie "A Beautiful Mind" once more, you will see that the meaning of a Nash equilibrium was not declared correctly. This was already noticed by S. Robinson in 2002. □

Remark 3.2. For further contributions of J. Nash to mathematics we refer to (59). □

4 The Poincaré-Miranda Theorem

The well-known intermediate-value theorem which says that a continuous function $f : [a, b] \to \mathbb{R}$ satisfying

$$f(a) \cdot f(b) \leq 0 \tag{4.1}$$

equals zero at some point of the interval $[a, b]$ was proved by Bernard Bolzano (1781-1848). In 1883-1884, Henri Poincaré announced the following generalization without proof:

> "Let $f_1, ..., f_n$ be n continuous functions of n variables $x_1, ..., x_n$: the variable x_i is subjected to vary between the limits a_i and $-a_i$. Let us suppose that for $x_i = a_i$, f_i is constantly positive, and that for $x_i = -a_i$, f_i is constantly negative; I say there will exist a system of values of x where the f's vanish."

In 1886 Poincaré published his argument on the homotopy invariance of the index which is a basis for the proof. The result obtained by Poincaré has come to be known as the theorem of Miranda, who in 1940 showed that it was equivalent to Brouwer's fixed point theorem. See (55).

4.1 The Theorem and its Corollaries

We will prove the Poincaré-Miranda theorem by using Brouwer's fixed point theorem as it was done by Miranda in 1940. The following corollaries will generalize the Poincaré-Miranda theorem.

Theorem 4.1. (The Poincaré-Miranda theorem) *Let* $\Omega = \{x \in \mathbb{R}^n : |x_i| \leq L_i,\ i = 1, ..., n\}$ *and* $f : \Omega \to \mathbb{R}^n$ *be continuous satisfying*

$$\begin{aligned} f_i(x_1, x_2, ..., x_{i-1}, -L_i, x_{i+1}, ..., x_n) \geq 0 \quad &\text{for} \quad 1 \leq i \leq n, \\ f_i(x_1, x_2, ..., x_{i-1}, +L_i, x_{i+1}, ..., x_n) \leq 0 \quad &\text{for} \quad 1 \leq i \leq n. \end{aligned} \tag{4.2}$$

Then, $f(x) = 0$ *has a solution in* Ω.

Proof: Case 1: $L_1 > 0, ..., L_n > 0$.

Case 1.a: Suppose that

$$\left.\begin{array}{l} f_i(x_1, .., x_{i-1}, -L_i, x_{i+1}, .., x_n) > 0 \\ f_i(x_1, .., x_{i-1}, +L_i, x_{i+1}, .., x_n) < 0 \end{array}\right\} i = 1, ..., n. \tag{4.3}$$

Since Ω is compact and since f is continuous, we can define for $i = 1, ..., n$

$$m_i := \min\{f_i(x) : x \in \Omega\} \quad \text{and} \quad M_i := \max\{f_i(x) : x \in \Omega\}.$$

By (4.3), we have $m_i < 0$ and $M_i > 0$. Now, we define

$$\delta_i^+ := \min\{|x_i + L_i| : x \in \Omega, f_i(x) < 0\}$$

and

$$\delta_i^- := \min\{|x_i - L_i| : x \in \Omega, f_i(x) > 0\}.$$

Thanks to (4.3), we have $\delta_i^+ > 0$ and $\delta_i^- > 0$ for all $i = 1, ..., n$. Because $m_i < 0$ and $M_i > 0$ for all $i = 1, ..., n$, there exist ε_i satisfying

$$0 < \varepsilon_i < \min\{-\frac{\delta_i^+}{m_i}, \frac{\delta_i^-}{M_i}\}, \quad i = 1, ..., n.$$

So, we can define $\tilde{f} : \Omega \to \mathbb{R}^n$ by

$$\tilde{f}_i(x) := x_i + \varepsilon_i \cdot f_i(x), \quad i = 1, ..., n.$$

We claim that

$$\tilde{f}(\Omega) \subseteq \Omega. \tag{4.4}$$

Let $i \in \{1, .., n\}$ be arbitrary but fixed. With $x \in \Omega$ we have $-L_i \leq x_i \leq L_i$. Firstly, if $f_i(x) = 0$, then

$$\tilde{f}_i(x) = x_i \in [-L_i, L_i].$$

Secondly, if $f_i(x) > 0$, then from

$$0 < f_i(x) \leq M_i \quad \text{and} \quad \varepsilon_i > 0$$

it follows

$$0 < \varepsilon_i \cdot f_i(x) \leq \varepsilon_i \cdot M_i$$

and together with $L_i - x_i \geq \delta_i^-$ one can conclude that

$$-L_i \leq x_i < \underbrace{x_i + \varepsilon_i \cdot f_i(x)}_{=\tilde{f}_i(x)} \leq L_i - \delta_i^- + \varepsilon_i \cdot M_i < L_i - \delta_i^- + \frac{\delta_i^-}{M_i} M_i = L_i.$$

Thirdly, if $f_i(x) < 0$, then from

$$0 > f_i(x) \geq m_i \quad \text{and} \quad \varepsilon_i > 0$$

it follows

$$0 > \varepsilon_i \cdot f_i(x) \geq \varepsilon_i \cdot m_i,$$

and together with $L_i + x_i \geq \delta_i^+$ one can conclude

$$L_i \geq x_i > \underbrace{x_i + \varepsilon_i \cdot f_i(x)}_{=\tilde{f}_i(x)} \geq -L_i + \delta_i^+ + \varepsilon_i \cdot m_i > -L_i + \delta_i^+ - \frac{\delta_i^+}{m_i} m_i = -L_i.$$

In other words, $\tilde{f}$ is a continuous self-mapping of Ω to Ω. Since Ω is nonempty, closed, bounded, and convex, we can apply Brouwer's fixed point theorem. This means that there exists some $x^* \in \Omega$ satisfying $\tilde{f}(x^*) = x^*$ which leads to $f(x^*) = 0$ due to the definition of $\tilde{f}$.

Case 1.b: Let (4.2) be true. Then, we define

$$f^{(k)}(x) = \left(f_1^{(k)}(x), ..., f_n^{(k)}(x)\right)^T$$

with

$$f_i^{(k)}(x) := f_i(x) - \frac{x_i}{k}, \quad i = 1, ..., n, \quad k = 1, 2, 3, ...$$

Since

$$f_i^{(k)}(x_1, .., x_{i-1}, -L_i, x_{i+1}, .., x_n) = f_i(x_1, .., x_{i-1}, -L_i, x_{i+1}, .., x_n) + \frac{L_i}{k} > 0$$

and

$$f_i^{(k)}(x_1, .., x_{i-1}, L_i, x_{i+1}, .., x_n) = f_i(x_1, .., x_{i-1}, L_i, x_{i+1}, .., x_n) - \frac{L_i}{k} < 0,$$

Case 1.a guarantees the existence of some $x^{(k)} \in \Omega$ satisfying $f^{(k)}(x^{(k)}) = 0$. Since Ω is compact, there exists a convergent subsequence $x^{(k_l)}$; i.e.,

$$\lim_{l \to \infty} x^{(k_l)} = x^* \quad \text{and} \quad x^* \in \Omega.$$

Due to the fact that f is continuous, we have

$$0 = \lim_{l \to \infty} f_i^{(k_l)}(x^{(k_l)}) = \lim_{l \to \infty} \left(f_i(x^{(k_l)}) - \frac{x_i^{(k_l)}}{k_l} \right) = f_i(x^*) \quad \text{for all } i = 1, .., n.$$

Case 2: $L_1 \geq 0, ..., L_n \geq 0$.

Case 2.a: $L_1 = 0, ..., L_n = 0$. Then, $\Omega = \{0\}$, and for all $i = 1, ..., n$ we have $f_i(0) \geq 0$ and $f_i(0) \leq 0$, which leads to $f(0) = 0$.

Case 2.b: Without loss of generality, we assume that

$$L_1 > 0, ..., L_k > 0,\ L_{k+1} = 0, ...,\ L_n = 0.$$

Following (4.2), we then have for $i = k+1, ..., n$

$$f_i(x_1, ..., x_k, 0, ..., 0) \geq 0 \quad \text{and} \quad f_i(x_1, ..., x_k, 0, ..., 0) \leq 0.$$

Hence, $f_i(x_1, ..., x_k, 0, ..., 0) = 0$ for $i = k+1, ..., n$. Defining $\tilde{\Omega} := \{x \in \mathbb{R}^k : |x_i| \leq L_i,\ i = 1, ..., k\}$ and $\tilde{f} : \tilde{\Omega} \to \mathbb{R}^k$ by

$$\tilde{f}_i(x) = f_i(x_1, ..., x_k, 0, ..., 0),\ i = 1, ..., k,$$

there exists

$$\begin{pmatrix} \tilde{x}_1 \\ \vdots \\ \tilde{x}_k \end{pmatrix} \in \tilde{\Omega}$$

satisfying $\tilde{f}(\tilde{x}^*) = 0$ according to Case 1. Finally, we have

$$f(\tilde{x}_1^*, ..., \tilde{x}_k^*, 0, ..., 0) = 0$$

with $(\tilde{x}_1^*, ..., \tilde{x}_k^*, 0, ..., 0)^{\mathrm{T}} \in \Omega$. □

The following corollary will show that it is not necessary that the center of the domain is the zero vector.

Corollary 4.1. *Let $x^c \in \mathbb{R}^n$, $L \in \mathbb{R}^n$, $L_i \geq 0$, $i = 1, ..., n$. Furthermore, let*

$$G = \{x = (x_i) \in \mathbb{R}^n : |x_i^c - x_i| \leq L_i,\ i = 1, .., n\},$$

$$\left.\begin{array}{l} G_i^+ = \{x \in G :\ x_i = x_i^c + L_i\} \\ G_i^- = \{x \in G :\ x_i = x_i^c - L_i\} \end{array}\right\} i = 1, .., n.$$

If $g : G \to \mathbb{R}$ is continuous satisfying

$$\left.\begin{array}{l} g_i(x) \leq 0 \text{ for all } x \in G_i^- \\ g_i(x) \geq 0 \text{ for all } x \in G_i^+ \end{array}\right\} i = 1, ..., n,$$

then there exists $x^* \in G$ satisfying $g(x^*) = 0$.

Proof: Considering $\Omega := \{\tilde{x} \in \mathbb{R}^n : |\tilde{x}_i| \leq L_i,\ i = 1,..,n\}$ and $f(\tilde{x}) = -g(x^c + \tilde{x})$ we have that $f : \Omega \to \mathbb{R}^n$ is continuous and for $\tilde{x} \in \Omega$ we get $f_i(\tilde{x}_1,..,\tilde{x}_{i-1}, -L_i, \tilde{x}_{i+1},..,\tilde{x}_n) =$

$$-g_i(\underbrace{x_1^c + \tilde{x}_1,..,x_{i-1}^c + \tilde{x}_{i-1}, x_i^c - L_i, x_{i+1}^c + \tilde{x}_{i+1},..,x_n^c + \tilde{x}_n}_{\in\ G_i^-}) \geq 0$$

and $f_i(\tilde{x}_1,..,\tilde{x}_{i-1}, L_i, \tilde{x}_{i+1},..,\tilde{x}_n) =$

$$-g_i(\underbrace{x_1^c + \tilde{x}_1,..,x_{i-1}^c + \tilde{x}_{i-1}, x_i^c + L_i, x_{i+1}^c + \tilde{x}_{i+1},..,x_n^c + \tilde{x}_n}_{\in\ G_i^+}) \leq 0.$$

According to Theorem 4.1 there exists $\tilde{x}^* \in \Omega$ with $f(\tilde{x}^*) = 0$. Defining $x^* := x^c + \tilde{x}^*$ we get $x^* \in G$ and $g(x^*) = g(x^c + \tilde{x}^*) = -f(\tilde{x}^*) = 0$. □

The following corollary is motivated by (4.1).

Corollary 4.2. *Let $x^c \in \mathbb{R}^n$, $L \in \mathbb{R}^n$, $L_i \geq 0$, $i = 1,...,n$ and $G = \{x = (x_i) \in \mathbb{R}^n : |x_i^c - x_i| \leq L_i, i = 1,..,n\}$. In addition, let*

$$\left.\begin{aligned} G_i^+ &= \{x \in G :\ x_i = x_i^c + L_i\} \\ G_i^- &= \{x \in G :\ x_i = x_i^c - L_i\} \end{aligned}\right\} i = 1,..,n.$$

If $g : G \to \mathbb{R}^n$ is continous satisfying

$$g_i(x) \cdot g_i(y) \leq 0 \text{ for all } x \in G_i^+ \text{ and for all } y \in G_i^-, i = 1,..,n, \tag{4.5}$$

then there exists some $x^* \in G$ with $g(x^*) = 0$.

Proof: Case 1: For all $i \in \{1,...,n\}$ either we have

$$g_i(x) \leq 0 \text{ for all } x \in G_i^- \text{ and } g_i(y) \geq 0 \text{ for all } y \in G_i^+ \tag{4.6}$$

or we have

$$g_i(x) \geq 0 \text{ for all } x \in G_i^- \text{ and } g_i(y) \leq 0 \text{ for all } y \in G_i^+. \tag{4.7}$$

In this case, we define $\tilde{g} : G \to \mathbb{R}^n$ by

$$\tilde{g}_i(x) = \begin{cases} g_i(x) \text{ if } i \text{ satisfies (4.6)} \\ -g_i(x) \text{ if } i \text{ satisfies (4.7).} \end{cases}$$

The set G and the function $\tilde{g}(x)$ fulfill the assumptions of Corollary 4.1. Therefore, there exists $x^* \in G$ with $\tilde{g}(x^*) = 0$, whence $g(x^*) = 0$.

Case 2: The set $\{1,...,n\}$ can be subdivided in two sets of indices, say I and J, so that for all $i \in I$ we have either (4.6) or (4.7), whereas for all $i \in J$ there exist either $x^{(1,i)}, x^{(2,i)} \in G_i^+$ or $x^{(1,i)}, x^{(2,i)} \in G_i^-$ satisfying $g_i(x^{(1,i)}) < 0$, $g_i(x^{(2,i)}) > 0$.

Without loss of generality, for all $i \in J$ let $x^{(1,i)}, x^{(2,i)} \in G_i^+$. Then, we can conclude from (4.5) that

$$\text{for all } i \in J \text{ we have} \quad g_i(y) = 0 \text{ for all } y \in G_i^- . \tag{4.8}$$

Again without loss of generality, we assume that

$$I = \{1, ..., n_0\} \text{ and } J = \{n_0 + 1, ..., n\}.$$

Then, we define $\tilde{G} := \{\tilde{x} \in \mathbb{R}^{n_0} : |x_i^c - \tilde{x}_i| \leq L_i,\ i = 1, ..., n_0\}$ and the function $\tilde{g} : \tilde{G} \to \mathbb{R}^{n_0}$ by

$$\tilde{g}(\tilde{x}) := \begin{pmatrix} g_1(\tilde{x}_1, ..., \tilde{x}_{n_0}, x_{n_0+1}^c - L_{n_0+1}, ..., x_n^c - L_n) \\ \vdots \\ g_{n_0}(\tilde{x}_1, ..., \tilde{x}_{n_0}, x_{n_0+1}^c - L_{n_0+1}, ..., x_n^c - L_n) \end{pmatrix}.$$

Due to Case 1, we can conclude that there exists some $\tilde{x}^* \in \tilde{G}$ with $\tilde{g}(\tilde{x}^*) = 0$. Setting

$$x^* := \begin{pmatrix} \tilde{x}_1^* \\ \vdots \\ \tilde{x}_{n_0}^* \\ x_{n_0+1}^c - L_{n_0+1} \\ \vdots \\ x_n^c - L_n \end{pmatrix}$$

we have $x^* \in G$ and $g(x^*) = 0$ by (4.8). □

4.2 A Fixed Point Version

Considering the original statement of Poincaré and considering the original paper of Miranda, one can see, that there is a tiny difference. It would be exactly the same, if (4.2) read

$$f_i(x_1, x_2, ..., x_{i-1}, -L_i, x_{i+1}, ..., x_n) < 0 \quad \text{for} \quad 1 \leq i \leq n,$$
$$f_i(x_1, x_2, ..., x_{i-1}, +L_i, x_{i+1}, ..., x_n) > 0 \quad \text{for} \quad 1 \leq i \leq n,$$

and if L_i was replaced by a_i. Of course, the $\leq$-sign is more general than the $<$-sign, but it is not clear, why Miranda changed

"... for $x_i = a_i$, f_i is constantly positive ..."

into

$$f_i(x_1, x_2, ..., x_{i-1}, +L_i, x_{i+1}, ..., x_n) \leq 0.$$

Of course, concerning the existence of a zero, this is irrelevant thanks Corollary 4.2. However, it is worth mentioning, that one also gets the existence of a fixed point, when using the original notation of Miranda. This will be shown in the following theorem.

Theorem 4.2. *Let* $\Omega = \{x \in \mathbb{R}^n : |x_i| \leq L_i, i = 1, ..., n\}$. *Furthermore, let* $g : \Omega \to \mathbb{R}^n$ *be continuous satisfying*

$$\left.\begin{aligned} g_i(x_1, ..., x_{i-1}, -L_i, x_{i+1}, ..., x_n) \geq 0 \\ g_i(x_1, ..., x_{i-1}, +L_i, x_{i+1}, ..., x_n) \leq 0 \end{aligned}\right\} \quad \text{for } i = 1, ..., n.$$

Then, there exists some $x^* \in \Omega$ *satisfying* $g(x^*) = x^*$.

Proof: Considering $f(x) := g(x) - x$, $x \in \Omega$ we have for all $i \in \{1, ..., n\}$

$$f_i(x_1, , .., x_{i-1}, -L_i, x_{i+1}, .., x_n) = g_i(x_1, .., x_{i-1}, -L_i, x_{i+1}, .., x_n) + L_i \geq 0,$$
$$f_i(x_1, , .., x_{i-1}, +L_i, x_{i+1}, .., x_n) = g_i(x_1, .., x_{i-1}, +L_i, x_{i+1}, .., x_n) - L_i \leq 0.$$

Then, from Theorem 4.1, we can conclude that there exists $x^* \in \Omega$ satisfying $f(x^*) = 0$. Therefore, $g(x^*) = x^*$. □

Analogous corollaries for this theorem (such as Corollary 4.1 and Corollary 4.2 for Theorem 4.1) are not true as we can see in Figure 4.1 and Figure 4.2, respectively.

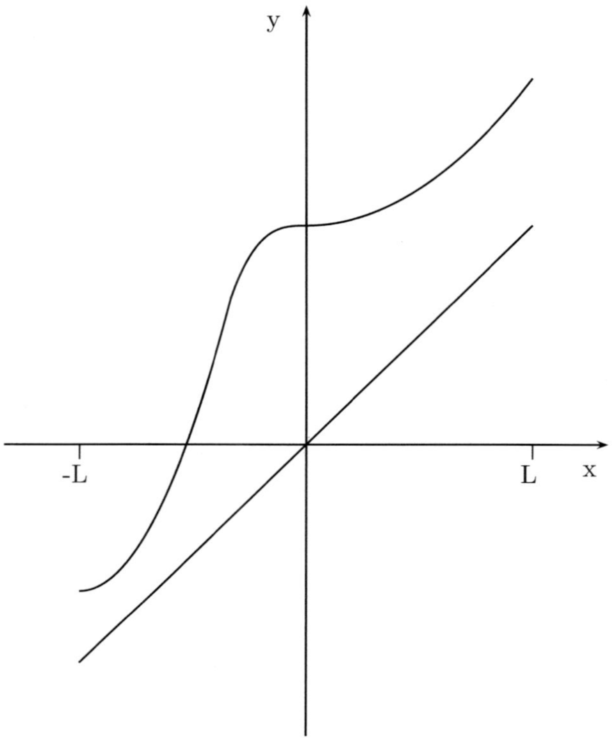

Figure 4.1: **No fixed point in** $[-L, L]$

In Figure 4.1 the graph of a function $y = g(x)$, $x \in [-L, L]$ is given satisfying $g(-L) < 0$, $g(L) > 0$. According to Corollary 4.2, $g(x)$ has a zero in $[-L, L]$. However, $g(x)$ has no fixed point in $[-L, L]$, which is no contradiction to Theorem 4.2, since $g(-L) \geq 0$, $g(L) \leq 0$ is not valid, here.

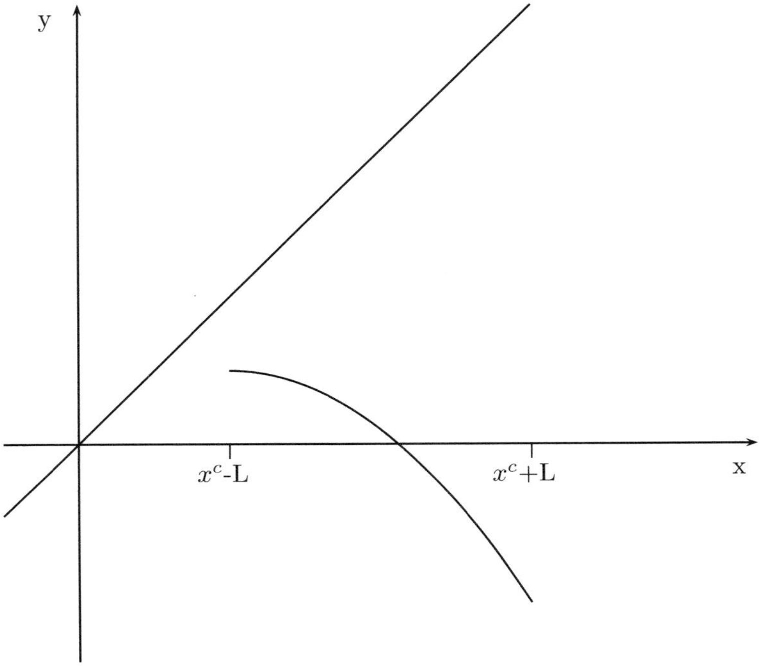

Figure 4.2: **No fixed point in** $[x^c - L, x^c + L]$**.**

In Figure 4.2 the graph of a function $y = g(x)$, $x \in [x^c - L, x^c + L]$ is given satisfying $g(x^c - L) > 0$, $g(x^c + L) < 0$. According to Corollary 4.1 $g(x)$ has a zero in $[x^c - L, x^c + L]$. However, $g(x)$ has no fixed point in $[x^c - L, x^c + L]$.

Remark 4.1. Note that the function g in Theorem 4.2 is not assumed to be a self-mapping as is assumed in many other fixed point theorems. □

5 The Nonlinear Complementarity Problem

Given a vector $f = (f_1, ..., f_n)^{\mathrm{T}}$ of n real, nonlinear functions of a real vector $z = (z_1, ..., z_n)^{\mathrm{T}}$, the nonlinear complementarity problem, abbreviated by NCP, is to find a vector z so that

$$z \geq 0, \quad f(z) \geq 0, \quad z^{\mathrm{T}} \cdot f(z) = 0,$$

or to show that no such vector exists. Here, the inequality sign is meant componentwise; i.e.,

$$\begin{aligned} z \geq 0 \quad &\Leftrightarrow \quad z_1 \geq 0, ..., z_n \geq 0 \\ f(z) \geq 0 \quad &\Leftrightarrow \quad f_1(z) \geq 0, ..., f_n(z) \geq 0. \end{aligned}$$

Especially, the condition $z^{\mathrm{T}} \cdot f(z) = 0$ then implies

$$z_i = 0 \quad \text{or} \quad f_i(z) = 0 \quad \text{for all } i = 1, ..., n.$$

In 1974, Tamir published an algorithm for solving the NCP for the case that f is a so-called Z-function. Tamir's algorithm is a generalization of Chandrasekaran's algorithm which solves the linear complementarity problem for the case that the given matrix M is a so-called Z-matrix.

Proving his results, Tamir used the iterative processes of Gauss-Seidel and of Jacobi. In Section 5.4 we present a different proof where these iterative processes are not used. Instead, we use the least element theory and the Poincaré-Miranda theorem.

Before we are going to show how the Poincaré-Miranda theorem can be used to prove the correctness of Tamir's algorithm, we give an application of the NCP where Tamir's algorithm can be used.

5.1 Free Boundary Problems

In contrast to a boundary problem where a differential equation has to be solved in a given domain satisfying given initial- and fixed boundary conditions, a free boundary problem is defined by a differential equation together with the unknown boundary of the domain.

A simple example is melting ice. If a piece of ice swims in a glass of water, the content of the glass is separated in two parts: the water and the ice. To determine the temperature of the water as a function dependent on time and space, one has to solve a parabolic differential equation (see (18)). The boundary of the water is moving since the ice is melting. Therefore, the boundary is not known from the start and it has to be determined as part of the solution.

For more examples concerning free boundrary problems we refer to the book of Crank. In the following example we show how the NCP arises from free boundary problems.

Example 5.1. We consider the ordinary free boundary problem:

$$\left.\begin{array}{c}\text{Find } s > 0 \text{ and } z(x) : [0, \infty) \to \mathbb{R} \text{ such that} \\ z''(x) = \sqrt{1 + (z(x))^2}, \quad \text{for } x \in [0, s], \\ z(0) = 1, \quad z'(s) = 0, \\ z(x) = 0, \quad \text{for } x \in [s, \infty). \end{array}\right\} \tag{5.1}$$

One can show that (5.1) has a unique solution, say $\{\hat{s}, \hat{z}(x)\}$, and that

$$\begin{array}{rcl} \hat{s} & \leq & \sqrt{2}, \\ \hat{z}(x) & > & 0, \quad x \in [0, \hat{s}). \end{array}$$

See (83), (99). However, it is not possible to solve (5.1) explicitly. So, one is satisfied to get approximations for $\hat{z}(x)$ at discrete points and to get an approximation for $\hat{s}$.

To show this, we choose $n \in \mathbb{N}$ and subdivide the interval $[0, \sqrt{2}]$ equidistantly; i.e., we define

$$h := \frac{\sqrt{2}}{n+1}, \quad x_i := i \cdot h, \quad i = 0, 1, ..., n+1.$$

Then, we have

$$\left.\begin{array}{ll} \hat{z}(x_i) > 0, \quad \text{if} & x_i \in [0, \hat{s}), \\ \hat{z}(x_i) = 0, \quad \text{if} & x_i \in [\hat{s}, \sqrt{2}]. \end{array}\right\} \tag{5.2}$$

For the case that $x_i \in [0, \hat{s}]$,

$$-\hat{z}''(x_i) + \sqrt{1 + \hat{z}(x_i)^2} = 0.$$

On the other hand, if $x_i \in (\hat{s}, \sqrt{2}]$, we get

$$-\hat{z}''(x_i) + \sqrt{1 + \hat{z}(x_i)^2} = -0 + \sqrt{1 + 0^2} = 1 > 0.$$

As a consequence, for all $i = 0, ..., n+1$ we have

$$\left.\begin{aligned} \hat{z}(x_i) &\geq 0, \\ -\hat{z}''(x_i) + \sqrt{1 + \hat{z}(x_i)^2} &\geq 0, \\ \hat{z}(x_i) \cdot \Big(- \hat{z}''(x_i) + \sqrt{1 + \hat{z}(x_i)^2} \Big) &= 0. \end{aligned}\right\} \tag{5.3}$$

Due to Taylor's formula with remainder, we get for $i = 1, ..., n$

$$\hat{z}(x_i + h) = \hat{z}(x_i) + h \cdot \hat{z}'(x_i) + \frac{1}{2}h^2 \cdot \hat{z}''(x_i) + \frac{1}{6}h^3 \cdot \hat{z}'''(x_i) + \frac{1}{24}h^4 \cdot \hat{z}''''(\xi_i)$$

and

$$\hat{z}(x_i - h) = \hat{z}(x_i) - h \cdot \hat{z}'(x_i) + \frac{1}{2}h^2 \cdot \hat{z}''(x_i) - \frac{1}{6}h^3 \cdot \hat{z}'''(x_i) + \frac{1}{24}h^4 \cdot \hat{z}''''(\nu_i)$$

with some $\nu_i \in (x_i - h, x_i)$ and some $\xi_i \in (x_i, x_i + h)$. Adding both equations we get

$$\frac{-\hat{z}(x_i - h) + 2 \cdot \hat{z}(x_i) - \hat{z}(x_i + h)}{h^2} = -\hat{z}''(x_i) - \frac{1}{24}h^2 \left(\hat{z}''''(\xi_i) + \hat{z}''''(\nu_i)\right).$$

So, for small h we have

$$\frac{-\hat{z}(x_i - h) + 2 \cdot \hat{z}(x_i) - \hat{z}(x_i + h)}{h^2} \approx -\hat{z}''(x_i).$$

Defining $z_0 := \hat{z}(0) = 1$ and $z_{n+1} := \hat{z}(\sqrt{2}) = 0$ we substitute

$$z_i \quad \text{for} \quad \hat{z}(x_i)$$

and

$$\frac{-z_{i-1} + 2 \cdot z_i - z_{i+1}}{h^2} \quad \text{for} \quad -\hat{z}''(x_i)$$

in (5.3); i.e., for $i = 1, ..., n$ we get

$$\left.\begin{aligned} z_i &\geq 0, \\ \frac{-z_{i-1} + 2 \cdot z_i - z_{i+1}}{h^2} + \sqrt{1 + z_i^2} &\geq 0, \\ z_i \cdot \left(\frac{-z_{i-1} + 2 \cdot z_i - z_{i+1}}{h^2} + \sqrt{1 + z_i^2} \right) &= 0. \end{aligned}\right\} \tag{5.4}$$

Defining the function $f : \mathbb{R}^n \to \mathbb{R}^n$ by

$$f(z) = \frac{1}{h^2}\begin{pmatrix} 2 & -1 & 0 & \cdots & 0 \\ -1 & 2 & -1 & \ddots & \vdots \\ 0 & \ddots & \ddots & \ddots & 0 \\ \vdots & \ddots & -1 & 2 & -1 \\ 0 & \cdots & 0 & -1 & 2 \end{pmatrix}\begin{pmatrix} z_1 \\ \vdots \\ \vdots \\ \vdots \\ z_n \end{pmatrix} + \begin{pmatrix} \sqrt{1+z_1^2} \\ \vdots \\ \vdots \\ \vdots \\ \sqrt{1+z_n^2} \end{pmatrix} + \begin{pmatrix} -\frac{1}{h^2} \\ 0 \\ \vdots \\ \vdots \\ 0 \end{pmatrix} \tag{5.5}$$

we see that (5.4) is an NCP.

This example will be continued at the end of this chapter. □

5.2 Least Element Theory

Tamir's algorithm presented in the next section was developed for the NCP defined by a so-called Z-function. So, we start with its definition.

Definition 5.1. *A function $f : \mathbb{R}^n \to \mathbb{R}^n$ is called a Z-function if for any $z \in \mathbb{R}^n$ the functions $\varphi_{ij}(t) := f_i(z + te_j)$, $i \neq j$, $i, j = 1, ..., n$ where e_j denotes the jth standard unit vector satisfy*

$$t \leq \tilde{t} \quad \Rightarrow \quad \varphi_{ij}(t) \geq \varphi_{ij}(\tilde{t}).$$

The function given in (5.5) is a Z-function. We'll see that Tamir's algorithm can be used to solve the discretized free boundary problem.

Moreover, Z-functions have been studied independently from free boundary problems. They have nice properties. One of them will be described in the following lemma.

Lemma 5.1. *Let $\mathbb{R}^k_+$ denote the first orthant of $\mathbb{R}^k$; i.e.,*

$$\mathbb{R}^k_+ = \{x \in \mathbb{R}^k : x_j \geq 0,\ j = 1, ..., k\}.$$

Furthermore, let $g : \mathbb{R}^k_+ \to \mathbb{R}^k$ be a Z-function and let

$$S := \{t \in \mathbb{R}^k_+ : g(t) \geq 0\}.$$

If $S \neq \emptyset$, then S has a least element.

Proof: Let $x, y \in S$ and $z := \inf\{x, y\}$, where

$$\inf\{x,y\} = \begin{pmatrix} \min\{x_1, y_1\} \\ \vdots \\ \min\{x_k, y_k\} \end{pmatrix}.$$

Then, since g is a Z-function and since $x, y \in S$ we have

$$\begin{aligned} x_i \le y_i \quad &\Rightarrow \quad g_i(z) \ge g_i(x) \ge 0, \\ y_i < x_i \quad &\Rightarrow \quad g_i(z) \ge g_i(y) \ge 0. \end{aligned}$$

Hence,

$$\inf\{x, y\} \in S. \tag{5.6}$$

Now, let $i \in \{1, ..., m\}$ be fixed but arbitrary. We define

$$M_i := \{s \in \mathbb{R} : \text{there exists } x \in S \text{ with } x_i = s\}.$$

Since S is bounded from below, M_i is bounded from below, too. Furthermore M_i is closed and is a nonempty subset of $\mathbb{R}$. Therefore, M_i has a least element, say m_i. So, we define

$$\tilde{x} = \begin{pmatrix} m_1 \\ \vdots \\ m_n \end{pmatrix}$$

and let $x^i \in S$ satisfy $x^i_i = m_i$. Then, we have

$$\tilde{x} = \inf\{x^1, ...x^n\} = \inf\{x^1, \inf\{x^2, ... \inf\{x^{n-1}, x^n\}...\}\}.$$

So we finally have $\tilde{x} \in S$ due to (5.6). □

Note, that not every bounded nonempty set $S \subset \mathbb{R}^n$ has a least element. See Figure 5.1.

We end this section with the definition of the infimum of a set.

Definition 5.2. *Let $S \subseteq \mathbb{R}^n$ be a nonempty set. Then, the greatest lower bound of S is denoted by* $\inf S$. *If $z := \inf S$, then it holds*

$$z \le t \quad \textit{for all } t \in S$$

and

$$\textit{if } v \le t \quad \textit{for all } t \in S \quad \textit{then } v \le z.$$

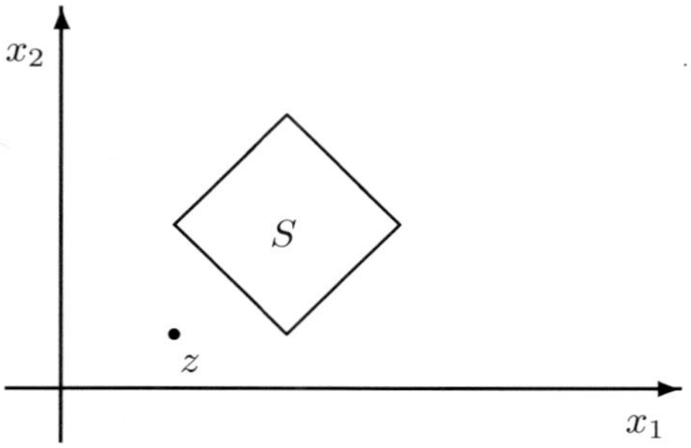

Figure 5.1: $z = \inf S \notin S$

5.3 Tamir's Algorithm

Tamir's algorithm is given in Table 5.1, where $\mathbb{R}^k_+$ denotes the first orthant of $\mathbb{R}^k$; i.e., $\mathbb{R}^k_+ = \{x \in \mathbb{R}^k : x_j \geq 0, j = 1, ..., k\}$. We want to remark that the pseudocode in Table 5.1 is not the original pseudocode presented by Tamir. We have removed the modified Jacobi process. Instead, we use the lines 5-7.

Example 5.2. (Wilhelm, 2008) Let

$$f(z) = \begin{pmatrix} 17 & -3 & -5 & -1 \\ -8 & 11 & -1 & -6 \\ -3 & -5 & 17 & -1 \\ -2 & -2 & -9 & 9 \end{pmatrix} \begin{pmatrix} z_1 \\ z_2 \\ z_3 \\ z_4 \end{pmatrix} + \begin{pmatrix} 22 + z_1^3 \\ -12 + z_2^3 \\ 15 + z_3^3 \\ -50 + z_4^3 \end{pmatrix}.$$

We will use Tamir's algorithm to solve the NCP defined by this Z-function. At the beginning, we have

$$k = 0, \quad z^{(0)} = 0, \quad f(z^{(0)}) = \begin{pmatrix} 22 \\ -12 \\ 15 \\ -50 \end{pmatrix} \ngeq 0.$$

We choose $i_1 = 2$ and get $J = \{2\}$. Now, Tamir's algorithm considers the function

$$g^{(1)} : \mathbb{R} \to \mathbb{R}$$
$$g^{(1)}(t) := f_2(0, t, 0, 0) = t^3 + 11t - 12.$$

begin
$k := 0\,;\ z := 0\,;\ J := \emptyset\,;$
if $f(z) \geq 0$ **then goto** 10

else repeat $k := k + 1;$
    choose $i_k \in \{1, ..., n\}$ with $f_{i_k}(z) < 0\,;$
    $J := J \cup \{i_k\}\,;$
    let $J = \{i_1, ..., i_k\}$ and $g^{(k)} : \mathbb{R}^k_+ \to \mathbb{R}^k$ be defined as

$$\begin{pmatrix} t_1 \\ \vdots \\ t_k \end{pmatrix} \mapsto \begin{pmatrix} f_{i_1}(\sum\limits_{j=1}^{k} t_j e_{i_j}) \\ \vdots \\ f_{i_k}(\sum\limits_{j=1}^{k} t_j e_{i_j}) \end{pmatrix};$$

5: let $M^{(k)} := \{t \in \mathbb{R}^k_+ : g^{(k)}(t) = 0,\ t_j \geq z_{i_j},\ j = 1, ..., k-1\};$
6: **if** $M^{(k)} \neq \emptyset$ **then**
7: **begin** $t^{(k)} := \inf M^{(k)};\ z := \sum\limits_{j=1}^{k} t_j^{(k)} e_{i_j}$ **end**
    else begin write('NCP(f) has no solution'); **goto** 20 **end**;
    until $f(z) \geq 0\,;$
10: **write**('The solution is ',z);
20: **end.**

Table 5.1: **Tamir's algorithm**

Note that $g^{(1)}(1) = 0$ and the function $g^{(1)}(t)$ is strictly increasing, since

$$\frac{\mathrm{d}}{\mathrm{dt}} g^{(1)}(t) = 3t^2 + 11 > 0.$$

This implies $M^{(1)} = \{1\}$. Going on with Tamir's algorithm, we get

$$t^{(1)} = 1, \quad z^{(1)} = \begin{pmatrix} 0 \\ 1 \\ 0 \\ 0 \end{pmatrix}, \quad \text{but } f(z^{(1)}) = \begin{pmatrix} 19 \\ 0 \\ 10 \\ -52 \end{pmatrix} \ngeq 0.$$

This leads to $i_2 = 4$ and $J = \{2, 4\}$. Now, Tamir's algorithm considers the function

$$g^{(2)} : \mathbb{R}^2 \to \mathbb{R}^2$$
$$g^{(2)}(t_1, t_2) := \begin{pmatrix} 11 & -6 \\ -2 & 9 \end{pmatrix} \begin{pmatrix} t_1 \\ t_2 \end{pmatrix} + \begin{pmatrix} -12 + t_1^3 \\ -50 + t_2^3 \end{pmatrix}.$$

Because of $g^{(2)}(2,3) = 0$, we have

$$\begin{pmatrix} 2 \\ 3 \end{pmatrix} \in \{t \in \mathbb{R}^2_+ : g^{(2)}(t) = 0,\, t_1 \geq 1\} = M^{(2)}.$$

Now, we want to show that $M^{(2)}$ is a singleton. So, assume that $g^{(2)}((a,b)) = 0$ with $a \geq 1$. Then,

$$\begin{aligned} a^3 + 11a - 6b - 12 &= 0, \\ b^3 + 9b - 2a - 50 &= 0. \end{aligned}$$

Substituting

$$b = \frac{a^3 + 11a - 12}{6} \tag{5.7}$$

in the second equation, we get

$$h(a) := \left(\frac{a^3 + 11a - 12}{6}\right)^3 + \frac{9}{6}(a^3 + 11a - 12) - 2a - 50 = 0.$$

Since

$$h'(a) = 3 \cdot \left(\frac{a^3 + 11a - 12}{6}\right)^2 \cdot \frac{3a^2 + 11}{6} + \frac{9}{2}a^2 + \frac{87}{6} > 0,$$

the function $h(a)$ is strictly increasing. So, there can be only one a with $h(a) = 0$. Then, via (5.7), also b is unique. As a result, $M^{(2)}$ is a singleton and we have $M^{(2)} = \{\begin{pmatrix} 2 \\ 3 \end{pmatrix}\}$. So, we define

$$t^{(2)} = \begin{pmatrix} 2 \\ 3 \end{pmatrix} \quad \text{and} \quad z^{(2)} = \begin{pmatrix} 0 \\ 2 \\ 0 \\ 3 \end{pmatrix}.$$

Since

$$f(z^{(2)}) = \begin{pmatrix} 13 \\ 0 \\ 2 \\ 0 \end{pmatrix} \geq 0,$$

the algorithm stops with a solution $z = \begin{pmatrix} 0 \\ 2 \\ 0 \\ 3 \end{pmatrix}$ of the NCP. □

5.4 The Correctness of Tamir's Algorithm

Before we prove the correctness of Tamir's algorthm, we need another lemma.

Lemma 5.2. *Let $f : \mathbb{R}^n_+ \to \mathbb{R}^n$ be a continuous Z-function and $M^{(k)}$ be defined as in Tamir's algorithm. If $M^{(k)} \neq \emptyset$, then* $\inf M^{(k)} \in M^{(k)}$.

Proof: We consider the sets $S^{(k)} := \{t \in \mathbb{R}^k_+ : g^{(k)}(t) \geq 0\}$. Since f is a Z-function, all $g^{(k)}$ are Z-functions, too. Since $M^{(k)} \subseteq S^{(k)}$, the condition $M^{(k)} \neq \emptyset$ implies $S^{(k)} \neq \emptyset$, and it follows by Lemma 5.1 that every $S^{(k)}$ has a least element, say $t^{(k)}$. First, we will show by induction, that $g^{(k)}(t^{(k)}) = 0$.

$k = 1$: With $f_{i_1}(0) < 0$ and $\emptyset \neq M^{(1)}$ the assertion follows by the intermediate-value theorem and the continuity of f.

$k - 1 \rightsquigarrow k$: Let $t^{(k-1)}$ be the least element of $S^{(k-1)}$, let

$$o = g^{(k-1)}(t^{(k-1)}) = \begin{pmatrix} f_{i_1}\left(\sum_{j=1}^{k-1} t_j^{(k-1)} e_{i_j}\right) \\ \vdots \\ f_{i_{k-1}}\left(\sum_{j=1}^{k-1} t_j^{(k-1)} e_{i_j}\right) \end{pmatrix}, \tag{5.8}$$

and let

$$0 > g_k^{(k)}\left(\begin{pmatrix} t^{(k-1)} \\ 0 \end{pmatrix}\right) = f_{i_k}\left(\sum_{j=1}^{k-1} t_j^{(k-1)} e_{i_j}\right). \tag{5.9}$$

By Lemma 5.1, $S^{(k)}$ has a least element $t^{(k)}$. Since $g^{(k-1)}$ is a Z-function, for all $j \in \{1, ..., k-1\}$

$$0 \leq f_{i_j}\left(\sum_{j=1}^{k} t_j^{(k)} e_{i_j}\right) \leq f_{i_j}\left(\sum_{j=1}^{k-1} t_j^{(k)} e_{i_j}\right).$$

Therefore, $(t_1^{(k)}, ..., t_{k-1}^{(k)})^{\mathrm{T}} \in S^{(k-1)}$, whence

$$\begin{pmatrix} t_1^{(k)} \\ \vdots \\ t_{k-1}^{(k)} \end{pmatrix} \geq \inf S^{(k-1)} = \begin{pmatrix} t_1^{(k-1)} \\ \vdots \\ t_{k-1}^{(k-1)} \end{pmatrix}. \tag{5.10}$$

Since $t^{(k)} \in S^{(k)}$, $g^{(k)}(t^{(k)}) \geq 0$. In order to prove $g^{(k)}(t^{(k)}) = 0$, we assume that there exists some $j \in \{1, ..., k\}$ with $g_j^{(k)}(t^{(k)}) > 0$. By (5.10), (5.8), and (5.9) we then have

$$t_j^{(k)} = \begin{pmatrix} t^{(k-1)} \\ 0 \end{pmatrix}_j \quad \Rightarrow \quad 0 < f_{i_j}\left(\sum_{j=1}^{k} t_j^{(k)} e_{i_j}\right) \leq f_{i_j}\left(\sum_{j=1}^{k-1} t_j^{(k-1)} e_{i_j}\right) \leq 0;$$

it follows that $t_j^{(k)} > \begin{pmatrix} t^{(k-1)} \\ 0 \end{pmatrix}_j$. So, we choose $\delta > 0$ so that

$$t_j^{(k)} - \delta > \begin{pmatrix} t^{(k-1)} \\ 0 \end{pmatrix}_j \quad \text{and} \quad g_j^{(k)}(t^{(k)} - \delta \cdot e_j) > 0.$$

This is possible, since $g^{(k)}$ is continuous. Then, we have

$$g_j^{(k)}(t^{(k)} - \delta \cdot e_j) > 0, \quad g_i^{(k)}(t^{(k)} - \delta \cdot e_j) \geq g_i^{(k)}(t^{k)}) \geq 0, \quad i \neq j.$$

Hence, $t^{(k)} - \delta \cdot e_j \in S^{(k)}$, which is a contradiction. Therefore, $g^{(k)}(t^{(k)}) = 0$. Together with (5.10) it follows $t^{(k)} \in M^{(k)}$.

Finally, we have to show $t^{(k)} = \inf M^{(k)}$. So, let $z := \inf M^{(k)}$. Since $t^{(k)} \in M^{(k)}$, it simply follows that $z \leq t^{(k)}$.

Next, since $t^{(k)} = \inf S^{(k)}$, we have

$$t^{(k)} \leq t \quad \text{for all } t \in S^{(k)}.$$

Because of $M^{(k)} \subseteq S^{(k)}$, we have

$$t^{(k)} \leq t \quad \text{for all } t \in M^{(k)}.$$

By the definition of the infimum, it holds

$$t^{(k)} \leq z.$$

So, we finally get $t^{(k)} = z = \inf M^{(k)}$. □

We will again state Corollary 4.1 because its notations can be used better within the proof in the following theorem.

Corollary 5.1. *Let $a, b \in \mathbb{R}^n$, $a \leq b$, Ω be the rectangle $\Omega := \{x \in \mathbb{R}^n : x_i \in [a_i, b_i],\ i = 1, ..., n\}$, and $f : \Omega \to \mathbb{R}^n$ be a continuous function on Ω. If*

$$\begin{aligned} f_i(x_1, x_2, ..., x_{i-1}, a_i, x_{i+1}, ..., x_n) \leq 0 \quad &\textit{for} \quad 1 \leq i \leq n, \\ f_i(x_1, x_2, ..., x_{i-1}, b_i, x_{i+1}, ..., x_n) \geq 0 \quad &\textit{for} \quad 1 \leq i \leq n, \end{aligned}$$

then $f(x) = 0$ has a solution in Ω.

Theorem 5.1. *Let $f : \mathbb{R}^n \to \mathbb{R}^n$ be a continuous Z-function. Then, Tamir's algorithm solves the NCP.*

Proof: We show by induction: If $M^{(k)} = \emptyset$, then the NCP has no solution.
$k = 1$: Let $f(0) \not\geq 0$, $f_{i_1}(0) < 0$, and $M^{(1)} = \emptyset$. Suppose, the NCP has a solution, say z^*. Then, we consider $g^{(1)}$ defined on the interval $[0, z^*_{i_1}]$. On the one hand, we have $g^{(1)}(0) = f_{i_1}(0) < 0$. On the other hand,

$$0 \leq f_{i_1}(z^*) = f_{i_1}\left(\sum_{j=1}^{n} z^*_j e_j\right) \leq f_{i_1}(z^*_{i_1} e_{i_1}) = g^{(1)}(z^*_{i_1}).$$

Since f is continuous, $g^{(1)}$ is continuous, too. So, by the intermediate-value theorem there must be some $\xi \in [0, z^*_{i_1}]$ satisfying $g^{(1)}(\xi) = 0$. This is a contradiction to the assumption that $M^{(1)} = \emptyset$ holds.
$k-1 \leadsto k$: Let $t^{(k-1)}$ be the least element of $S^{(k-1)}$ satisfying (5.8) and (5.9), and let $M^{(k)} = \emptyset$. Suppose, the NCP has a solution, say z^*. Then,

$$0 \leq f_{i_j}(z^*) = f_{i_j}\left(\sum_{j=1}^{n} z^*_j e_j\right) \leq f_{i_j}\left(\sum_{j=1}^{k-1} z^*_{i_j} e_{i_j}\right), \quad j = 1, ..., k-1. \tag{5.11}$$

Hence, $(z^*_{i_1}, ..., z^*_{i_{k-1}})^{\mathrm{T}} \in S^{(k-1)}$ and it follows that $(z^*_{i_1}, ..., z^*_{i_{k-1}})^{\mathrm{T}} \geq t^{(k-1)}$. Now, we consider the function $g^{(k)}$ on the product of intervals

$$\Omega = \begin{pmatrix} [t^{(k-1)}_1, z^*_{i_1}] \\ \vdots \\ [t^{(k-1)}_{k-1}, z^*_{i_{k-1}}] \\ [0, z^*_{i_k}] \end{pmatrix}.$$

Let $t \in \Omega$; i.e., $t_j \in [t^{(k-1)}_j, z^*_{i_j}]$, $j = 1, ..., k-1$ and $t_k \in [0, z^*_{i_k}]$. Then, for all $l \in \{1, ..., k-1\}$ we have

$$f_{i_l}\left(\sum_{j=1,\, j\neq l}^{k-1} t_j e_{i_j} + t^{(k-1)}_l e_{i_l} + t_k e_{i_k}\right) \leq f_{i_l}\left(\sum_{j=1}^{k-1} t^{(k-1)}_j e_{i_j}\right) = 0$$

by (5.8) and

$$f_{i_l}\left(\sum_{j=1,\, j\neq l}^{k-1} t_j e_{i_j} + z^*_{i_l} e_{i_l} + t_k e_{i_k}\right) \geq f_{i_l}\left(\sum_{j=1}^{k} z^*_{i_j} e_{i_j}\right) \geq f_{i_l}(z^*) \geq 0.$$

Finally,

$$f_{i_k}\left(\sum_{j=1}^{k-1} t_j e_{i_j} + 0 \cdot e_{i_k}\right) \leq f_{i_k}\left(\sum_{j=1}^{k-1} t_j^{(k-1)} e_{i_j}\right) < 0$$

by (5.9) and

$$f_{i_k}\left(\sum_{j=1}^{k-1} t_j e_{i_j} + z_{i_k}^* \cdot e_{i_k}\right) \geq f_{i_k}(z^*) \geq 0.$$

Since f is continuous, $g^{(k)}$ is continuous, too. So, by Corollary 5.1 applied to $g^{(k)}$ and Ω there must be some $\xi \in \Omega$ satisfying $g^{(k)}(\xi) = 0$. This is a contradiction to the assumption that $M^{(k)} = \emptyset$ holds. □

Example 5.3. We continue Example 5.1. F. Wilhelm has shown that $f(z)$, $z \geq 0$ is injective. As a result, applying Tamir's algorithm for solving the NCP, all sets $M^{(k)}$ are either empty or a singleton. In contrast to the original paper of Tamir, the method for calculating a zero of $g^{(k)}$ is not fixed in Table 5.1. So, it is left to the programmer which method for calculating a zero is chosen.

The results presented in Table 5.2 are based on an implementation of D. Hammer. The input data are n as well as the tolerance $\varepsilon > 0$. As the method for calculating a zero of $g^{(k)}$, Newton's method was chosen, where

$$t_{start} := \begin{cases} 0 & \text{if } k = 1 \\ \begin{pmatrix} t^{(k-1)} \\ 0 \end{pmatrix} & \text{if } k > 1 \end{cases}$$

was taken as the starting point, respectively. If $z_i = 0$ and $z_{i+1} > 0$, then $\tilde{s} := \frac{1}{2}(x_i + x_{i+1})$ was taken as an approximation for $\hat{s}$. See Table 5.2 for some examples. Note, that the exact value of $\hat{s}$ satisfies $\hat{s} \in [1.393206, 1.397715]$; see (84). □

n	$\tilde{s}$	running time
50	1.372619	0.017 s
100	1.379208	0.114 s
150	1.390799	0.720 s
200	1.389587	1.507 s
250	1.388859	3.962 s
500	1.387397	20.478 s

n	$\tilde{s}$	running time
50	1.372619	0.028 s
100	1.393210	0.201 s
150	1.390799	0.831 s
200	1.389587	2.192 s
250	1.388859	4.577 s
500	1.393042	29.514 s

Table 5.2: $\varepsilon = 10^{-5}$ $\qquad$ $\varepsilon = 10^{-11}$

6 Verification Methods

Many numerical methods for calculating a zero of a function $f : \mathbb{R}^n \to \mathbb{R}^n$ involve calculating an approximate solution, say $\tilde{x}$. These methods often consist of iterative algorithms using the stopping criterion

$$\|f(\tilde{x})\|_\infty < \varepsilon, \tag{6.1}$$

where, here, $\|f(\tilde{x})\|_\infty := \max\{|f_1(\tilde{x})|, ..., |f_n(\tilde{x})|\}$ and where $\varepsilon > 0$ is a given tolerance. However, the criterion (6.1) is not sufficient in order to conclude that there is a solution of $f(x) = 0$.

Example 6.1. Let $f : \mathbb{R}^2 \to \mathbb{R}^2$ defined by

$$f(x, y) = \begin{pmatrix} f_1(x, y) \\ f_2(x, y) \end{pmatrix} = \begin{pmatrix} \min\{x, -\frac{1}{2}x + \frac{1}{2}y - \varepsilon\} \\ \min\{y, -\frac{5}{2}x + \frac{5}{2}y - 3\varepsilon\} \end{pmatrix}.$$

Assume, an approximate solution is given by

$$(\tilde{x}, \tilde{y}) = (0, \varepsilon).$$

Then, $\|f(\tilde{x}, \tilde{y})\|_\infty = \frac{1}{2}\varepsilon < \varepsilon$. However, a zero of f does not exist. Assume, $f(x^*, y^*) = (0, 0)$. Then, the first component gives

$$x^* = 0 \quad \text{or} \quad -\frac{1}{2}x^* + \frac{1}{2}y^* - \varepsilon = 0.$$

Case 1: Let $x^* = 0$. Then, $y^* \geq 2\varepsilon > 0$. Substituting this result in $f_2(x^*, y^*) = 0$, we get a contradiction; after all,

$$0 = f_2(x^*, y^*) = \min\{y^*, -\frac{5}{2} \cdot 0 + \frac{5}{2} \cdot y^* - 3\varepsilon\} \geq \min\{2\varepsilon, \frac{5}{2} \cdot 2\varepsilon - 3\varepsilon\} = 2\varepsilon > 0.$$

Case 2: Let $-\frac{1}{2}x^* + \frac{1}{2}y^* - \varepsilon = 0$ (and $x^* > 0$). It follows

$$y^* = x^* + 2\varepsilon > 0.$$

Substituting this result within $f_2(x^*, y^*) = 0$, again we get a contradiction

$$\begin{aligned} 0 = f_2(x^*, y^*) &= \min\{y^*, -\tfrac{5}{2}x^* + \tfrac{5}{2}y^* - 3\varepsilon\} \\ &= \min\{x^* + 2\varepsilon, -\tfrac{5}{2}x^* + \tfrac{5}{2} \cdot (x^* + 2\varepsilon) - 3\varepsilon\} = 2\varepsilon > 0. \quad \square \end{aligned}$$

Considering (6.1), we only can *guess* that in the so-called interval vector

$$\begin{pmatrix} [\tilde{x}_1 - \varepsilon, \tilde{x}_1 + \varepsilon] \\ \vdots \\ [\tilde{x}_n - \varepsilon, \tilde{x}_n + \varepsilon] \end{pmatrix}$$

there exists an x^* satisfying $f(x^*) = 0$. A method done by use of a computer that verifies (i.e. that proves) that this guess is really true is called a verification method.

How can this be done?

The basic idea is very simple: Check by use of a computer that (4.5) is valid for

$$\Omega = \begin{pmatrix} [\tilde{x}_1 - \varepsilon, \tilde{x}_1 + \varepsilon] \\ \vdots \\ [\tilde{x}_n - \varepsilon, \tilde{x}_n + \varepsilon] \end{pmatrix}.$$

Then, the corollaries of the Poincaré-Miranda theorem prove that there is some $x^* \in \Omega$ with $f(x^*) = 0$.

Another idea is to transform the problem $f(x) = 0$ into a fixed point problem and to verify by use of a computer the assumptions of Brouwer's fixed point theorem.

We present three verification methods. One of them is based on Brouwer's fixed point theorem and two of them are based on the Poincaré-Miranda theorem. All of them use interval computations.

6.1 Interval Computations

Let $[a]$ denote a real compact interval; i.e.,

$$[a] = [\underline{a}, \overline{a}] = \{x \in \mathbb{R} : \underline{a} \le x \le \overline{a}\}.$$

Sometimes we also write

$$\sup[a] \text{ instead of } \overline{a} \quad \text{and} \quad \inf[a] \text{ instead of } \underline{a}.$$

Especially we do this in the case that the interval is described by a long term, see Section 6.6.

Furthermore, let **IR** denote the set of all real compact intervals. Then, we can define $+, -, \cdot, /$ in the set theoretic sense; i.e.,

$$\begin{aligned}
[a]+[b] &= \{a+b : a \in [a],\, b \in [b]\},\\
[a]-[b] &= \{a-b : a \in [a],\, b \in [b]\},\\
[a]\cdot[b] &= \{a\cdot b : a \in [a],\, b \in [b]\},\\
\frac{[a]}{[b]} &= \{\frac{a}{b} : a \in [a],\, b \in [b]\} \quad \text{if } 0 \notin [b].
\end{aligned}$$

Example 6.2. Let $[a] = [1,3]$ and $[b] = [2,4]$. Then,

$$\begin{aligned}
[a]+[b] &= \{a+b : a \in [1,3],\, b \in [2,4]\} = [3,7],\\
[a]-[b] &= \{a-b : a \in [1,3],\, b \in [2,4]\} = [-3,1],\\
[a]\cdot[b] &= \{a\cdot b : a \in [1,3],\, b \in [2,4]\} = [2,12],\\
\frac{[a]}{[b]} &= \{\frac{a}{b} : a \in [1,3],\, b \in [2,4]\} = [\frac{1}{4},\frac{3}{2}].
\end{aligned}$$

□

Following Example 6.2, it is easy to see that the results of $[a] * [b]$, where $* \in \{+,-,\cdot,/\}$ can be expressed by the interval boundaries of $[a]$ and $[b]$. More precisely,

$$\left.\begin{aligned}
[a]+[b] &= [\underline{a}+\underline{b}, \overline{a}+\overline{b}],\\
[a]-[b] &= [\underline{a}-\overline{b}, \overline{a}-\underline{b}],\\
[a]\cdot[b] &= [\min\{\underline{a}\cdot\underline{b}, \underline{a}\cdot\overline{b}, \overline{a}\cdot\underline{b}, \overline{a}\cdot\overline{b}\}, \max\{\underline{a}\cdot\underline{b}, \underline{a}\cdot\overline{b}, \overline{a}\cdot\underline{b}, \overline{a}\cdot\overline{b}\}]\\
\frac{[a]}{[b]} &= [a]\cdot[\frac{1}{\overline{b}},\frac{1}{\underline{b}}] \quad \text{if } 0 \notin [b].
\end{aligned}\right\} \qquad (6.2)$$

So, for any $a \in [a]$ and any $b \in [b]$ we have

$$a * b \in [a] * [b], \quad * \in \{+,-,\cdot,/\}.$$

(For the case $* = /$ we have to assume, that $0 \notin [b]$). This means, that any term $a * b$ is included within the interval $[a] * [b]$. For example,

$$\frac{1}{3} + \pi \in [0.333, 0.334] + [3.141, 3.142] = [3.474, 3.476].$$

The smaller the interval will be, the better an approximation of the exact value can be propagated. But interval arithmetic is not as easy as it seems. Some rules, that you are used to know in $\mathbb{R}$, are not valid in **IR**. For example, the distributivity law, which reads

$$a\cdot(b+c) = a\cdot b + a\cdot c,$$

is not valid in **IR**. Let $[a] = [1, 2]$, $[b] = [-1, 1]$, and $[c] = [2, 4]$. Then,

$$[a] \cdot ([b] + [c]) = [1, 2] \cdot [1, 5] = [1, 10],$$

whereas

$$[a] \cdot [b] + [a] \cdot [c] = [-2, 2] + [2, 8] = [0, 10].$$

One can show, that only the so-called sub-distributivity law

$$[a] \cdot ([b] + [c]) \subseteq [a] \cdot [b] + [a] \cdot [c]$$

is valid in **IR**. We refer to the book of G. Alefeld and J. Herzberger for a detailed introduction into interval computations.

The main thing, that we need in the following sections, is the meaning of a so-called interval evaluation.

Example 6.3. Let $f(x) = x^3 + 2x^2 + x + 2$ and let $[x] = [1, 2]$. Then, an interval evaluation is defined via the basic rules (6.2) as follows:

$$\begin{aligned} f([x]) &= f([1, 2]) = [1, 2] \cdot [1, 2] \cdot [1, 2] + 2 \cdot [1, 2] \cdot [1, 2] + [1, 2] + 2 \\ &= [1, 8] + [2, 8] + [3, 4] = [6, 20]. \end{aligned}$$

Clearly, if there is some unknown number, say η, that we can only include in an interval, say $\eta \in [1, 2]$, then we have $f(\eta) \in f([1, 2])$. □

To get a feeling where we can really benefit from an interval evaluation, we consider the following detailed example.

Example 6.4. We consider the boundary value problem:

$$\left.\begin{aligned} -y''(x) &= f(x, y(x), y'(x)), \quad x \in [a, b] \\ y(a) &= \alpha, \\ y(b) &= \beta. \end{aligned}\right\} \tag{6.3}$$

To present the main ideas, we just consider a special example.

$$\left.\begin{aligned} -y''(x) &= e^x, \quad x \in [0, 1] \\ y(0) &= 0, \\ y(1) &= 0. \end{aligned}\right\} \tag{6.4}$$

We choose $n \in \mathbb{N}$ and subdivide the interval $[0, 1]$ equidistantly; i.e., we define

$$h := \frac{1}{n+1}, \quad x_i := i \cdot h, \quad i = 0, 1, ..., n + 1.$$

Let $\hat{y}(\cdot)$ denote the solution, then Taylor's formula with remainder term leads to

$$\hat{y}(x_i+h)=\hat{y}(x_i)+h\cdot\hat{y}'(x_i)+\frac{1}{2}h^2\cdot\hat{y}''(x_i)+\frac{1}{6}h^3\cdot\hat{y}'''(x_i)+\frac{1}{24}h^4\cdot\hat{y}''''(\xi_i)$$

and

$$\hat{y}(x_i-h)=\hat{y}(x_i)-h\cdot\hat{y}'(x_i)+\frac{1}{2}h^2\cdot\hat{y}''(x_i)-\frac{1}{6}h^3\cdot\hat{y}'''(x_i)+\frac{1}{24}h^4\cdot\hat{y}''''(\nu_i)$$

with some $\nu_i\in(x_i-h,x_i)$ and some $\xi_i\in(x_i,x_i+h)$. Adding both equations we get

$$\frac{-\hat{y}(x_i-h)+2\cdot\hat{y}(x_i)-\hat{y}(x_i+h)}{h^2}=-\hat{y}''(x_i)-\frac{h^2}{24}\left(\hat{y}''''(\xi_i)+\hat{y}''''(\nu_i)\right)\quad(6.5)$$

for $i=1,...,n$. Keeping in mind, that $-\hat{y}''(x)=e^x$, we know $\hat{y}''''(x)=-e^x$, too. For example, let $n=1$. Then, $h=\frac{1}{2}$, $x_0=0$, $x_1=\frac{1}{2}$, $x_2=1$ and

$$\nu_1\in[x_1-h,x_1]=[0,\frac{1}{2}],\quad \xi_1\in[x_1,x_1+h]=[\frac{1}{2},1].$$

Formula (6.5) then reads

$$\frac{-\hat{y}(0)+2\cdot\hat{y}(\frac{1}{2})-\hat{y}(1)}{\left(\frac{1}{2}\right)^2}=e^{\frac{1}{2}}+\frac{1}{24}\left(\frac{1}{2}\right)^2\left(e^{\xi_1}+e^{\nu_1}\right).$$

Using $\hat{y}(0)=0$, $\hat{y}(1)=0$, $\nu_1\in[0,\frac{1}{2}]$, and $\xi_1\in[\frac{1}{2},1]$, the interval evaluation gives

$$\begin{aligned}\hat{y}(\tfrac{1}{2}) &\in \frac{1}{8}e^{\frac{1}{2}}+\frac{1}{8}\cdot\frac{1}{96}\left(e^{[0,\frac{1}{2}]}+e^{[\frac{1}{2},1]}\right)\\ &= \frac{1}{8}e^{\frac{1}{2}}+\frac{1}{8}\cdot\frac{1}{96}\cdot[1+e^{\frac{1}{2}},e^{\frac{1}{2}}+e]\subseteq[0.209539,0.211776].\end{aligned}$$

For $n>1$ formula (6.5) leads to

$$\frac{1}{h^2}\begin{pmatrix}2 & -1 & 0 & \cdots & 0\\ -1 & 2 & -1 & \ddots & \vdots\\ 0 & \ddots & \ddots & \ddots & 0\\ \vdots & \ddots & -1 & 2 & -1\\ 0 & \cdots & 0 & -1 & 2\end{pmatrix}\cdot\begin{pmatrix}\hat{y}(x_1)\\ \vdots\\ \vdots\\ \vdots\\ \hat{y}(x_n)\end{pmatrix}=\begin{pmatrix}e^{x_1}+\frac{1}{24}\cdot h^2(e^{\xi_1}+e^{\nu_1})\\ \vdots\\ \vdots\\ \vdots\\ e^{x_n}+\frac{1}{24}\cdot h^2(e^{\xi_n}+e^{\nu_n})\end{pmatrix}$$

with some $\nu_i\in(x_i-h,x_i)$ and some $\xi_i\in(x_i,x_i+h)$, $i=1,...,n$.

Applying the interval Cholesky method (see (5)) we get for $n = 8$:

$$\begin{aligned}
\hat{y}\left(\frac{1}{9}\right) &\in [0.073376707, 0.073463768] \\
\hat{y}\left(\frac{2}{9}\right) &\in [0.132943708, 0.133096479] \\
\hat{y}\left(\frac{3}{9}\right) &\in [0.177080357, 0.177276777] \\
\hat{y}\left(\frac{4}{9}\right) &\in [0.203974522, 0.204192766] \\
\hat{y}\left(\frac{5}{9}\right) &\in [0.211601374, 0.211819618] \\
\hat{y}\left(\frac{6}{9}\right) &\in [0.197698130, 0.197894550] \\
\hat{y}\left(\frac{7}{9}\right) &\in [0.159736085, 0.159888857] \\
\hat{y}\left(\frac{8}{9}\right) &\in [0.094889365, 0.094976663].
\end{aligned}$$

The problem (6.4) was chosen, so that the solution can be expressed explicitly in order to compare the results. The solution of (6.4) reads

$$\hat{y}(x) = -e^x + (e-1) \cdot x + 1.$$

So, we have $\hat{y}\left(\frac{1}{2}\right) = 0.21042...$ and

$$\begin{aligned}
\hat{y}\left(\frac{1}{9}\right) &= 0.073401... \\
\hat{y}\left(\frac{2}{9}\right) &= 0.132992... \\
\hat{y}\left(\frac{3}{9}\right) &= 0.177148... \\
\hat{y}\left(\frac{4}{9}\right) &= 0.204057... \\
\hat{y}\left(\frac{5}{9}\right) &= 0.211692... \\
\hat{y}\left(\frac{6}{9}\right) &= 0.197787... \\
\hat{y}\left(\frac{7}{9}\right) &= 0.159811... \\
\hat{y}\left(\frac{8}{9}\right) &= 0.094936....
\end{aligned}$$

The idea to use Taylor's formula with remainder term and to use interval computations in order to get inclusions of the exact solution at discrete points, was presented by E. Hansen and R. Moore. It was applied to problem (6.3), where the existence of a unique solution is given, but where this solution cannot be stated explicitly as in our simple example (6.4). For the technical details we refer to (43), for example. □

Interval computations can easily be extended to the n-dimensional case. An interval matrix, for instance, is a matrix with an interval in each entry; i.e.,

$$[A] = \begin{pmatrix} [1,2] & [2,3] \\ [-1,0] & [1,3] \end{pmatrix}.$$

By $\mathbf{IR}^{m \times n}$, we denote the set of all interval matrices with m columns and n rows. Then, with $[A]$, $[B] \in \mathbf{IR}^{n \times n}$, we define

$$[A] \pm [B] = ([a_{ij}]) \pm ([b_{ij}]) = ([a_{ij}] \pm [b_{ij}])$$

and

$$[A] \cdot [B] = (\sum_{k=1}^{n} [a_{ik}] \cdot [b_{kj}]),$$

where the computations of intervals are done as defined in (6.2).

An interval evaluation can also be defined as in the 1-dimensional case.

Example 6.5. Let

$$f(x,y) = x \cdot y^2, \quad x \in [-2,1], \, y \in [1,2].$$

Then, $f([x],[y]) = [-2,1] \cdot [1,2]^2 = [-2,1] \cdot [1,4] = [-8,4]$. □

Finally, we want to note that not every function has an interval evaluation.

Example 6.6. Consider the function

$$f(x) = \begin{cases} \dfrac{\sin x}{x}, & x \neq 0, \\ 1, & x = 0. \end{cases}$$

Then, $f([-1,1])$ is not defined. □

6.2 The Krawczyk Operator

To motivate the definition of the Krawczyk operator, we consider a differentiable function $f : \mathbb{R} \to \mathbb{R}$. The mean-value theorem gives

$$f(y) - f(x^{(0)}) = f'(\xi) \cdot (y - x^{(0)}) \text{ for some } \xi \text{ between } x^{(0)} \text{ and } y. \qquad (6.6)$$

If $f(y) = 0$, then

$$0 = f(x^{(0)}) + f'(\xi) \cdot (y - x^{(0)}). \qquad (6.7)$$

Multiplying (6.7) by some $r \neq 0$, we get

$$0 = r \cdot f(x^{(0)}) + r \cdot f'(\xi) \cdot (y - x^{(0)}). \qquad (6.8)$$

In order to transform (6.8) into a fixed point equation, we add and subtract the term $y - x^{(0)}$, getting

$$0 = r \cdot f(x^{(0)}) + (r \cdot f'(\xi) - 1) \cdot (y - x^{(0)}) + y - x^{(0)}.$$

Thus, $y = x^{(0)} - r \cdot f(x^{(0)}) + (1 - r \cdot f'(\xi)) \cdot (y - x^{(0)})$. Now, let $[x] \in \mathbf{IR}$ be an interval with $y \in [x]$ and $x \in [x]$. Then, also $\xi \in [x]$. Suppose the interval evaluation of $f'([x])$ is possible as defined in Section 6.1. Then, $f'(\xi) \in f'([x])$. The Krawczyk operator is then defined as

$$K(x^{(0)}, r, [x]) = x^{(0)} - r \cdot f(x^{(0)}) + (1 - r \cdot f'([x])) \cdot ([x] - x^{(0)}),$$

where the interval computations are defined as in Section 6.1.

It is often useful to have a multi-dimensional generalization of the Krawczyk operator along with a corresponding generalized method of interval computation. The starting point above was the mean-value theorem. This theorem is only valid in the 1-dimensional case. However, the Krawczyk operator can also be defined for functions $f : \mathbb{R}^n \to \mathbb{R}^n$ as follows.

Again, we assume that f is differentiable. Then,

$$f(y) - f(x^{(0)}) = B(x^{(0)}, y) \cdot (y - x^{(0)}), \qquad (6.9)$$

where

$$B(x^{(0)}, y) = \begin{pmatrix} \frac{\partial f_1}{\partial x_1}\big(x^{(0)} + \lambda_1 \cdot \big(y - x^{(0)}\big)\big) & \cdots & \frac{\partial f_1}{\partial x_n}\big(x^{(0)} + \lambda_1 \cdot \big(y - x^{(0)}\big)\big) \\ \vdots & & \vdots \\ \frac{\partial f_n}{\partial x_1}\big(x^{(0)} + \lambda_n \cdot \big(y - x^{(0)}\big)\big) & \cdots & \frac{\partial f_n}{\partial x_n}\big(x^{(0)} + \lambda_n \cdot \big(y - x^{(0)}\big)\big) \end{pmatrix}$$

with some $\lambda_1, ..., \lambda_n \in (0, 1)$. [1] Note that, in general, the λ_i will all be distinct.

[1] See Section 3.2 in (68).

Now, let $R \in \mathbb{R}^{n\times n}$ be a nonsingular matrix, let I denote the unit matrix, let $[x] \in \mathbf{IR}^{n\times n}$ be an interval vector, let $y \in [x]$, and let $x^{(0)} \in [x]$ be arbitrary but fixed. If $\frac{\partial f_i}{\partial x_j}([x])$ can be evaluated for $i, j = 1, ..., n$ as described in Section 6.1, then we have

$$B(x^{(0)}, y) \in f'([x]),$$

where

$$f'([x]) = \begin{pmatrix} \frac{\partial f_1}{\partial x_1}([x]) & \cdots & \frac{\partial f_1}{\partial x_n}([x]) \\ \vdots & & \vdots \\ \frac{\partial f_n}{\partial x_1}([x]) & \cdots & \frac{\partial f_n}{\partial x_n}([x]) \end{pmatrix},$$

because $x^{(0)} + \lambda_i \cdot \left(y - x^{(0)}\right) \in [x]$ for $i = 1, ..., n$, since $[x]$ is convex. As a result, the Krawczyk operator can be defined as

$$K(x^{(0)}, R, [x]) = x^{(0)} - R \cdot f(x^{(0)}) + (I - R \cdot f'([x])) \cdot ([x] - x^{(0)}).$$

Another technique for finding a matrix $B(x^{(0)}, y)$ satiyfying (6.9) goes back to E. Hansen. For this, it is sufficient that the function f is continuous. Because of

$$f_i(y_1, \ldots, y_n) - f_i\left(x_1^{(0)}, \ldots, x_n^{(0)}\right) =$$

$$f_i(y_1, \ldots, y_n) - f_i\left(x_1^{(0)}, y_{2,} \ldots, y_n\right) + f_i\left(x_1^{(0)}, y_{2,} \ldots, y_n\right)$$

$$-f_i\left(x_1^{(0)}, x_2^{(0)}, y_3, \ldots, y_n\right) + f_i\left(x_1^{(0)}, x_2^{(0)}, y_3, \ldots, y_n\right)$$

$$\pm \cdots - f_i\left(x_1^{(0)}, \ldots, x_{n-1}^{(0)}, y_n\right) + f_i\left(x_1^{(0)}, \ldots, x_{n-1}^{(0)}, y_n\right) - f_i\left(x_1^{(0)}, \ldots, x_n^{(0)}\right) =$$

$$\sum_{j=1}^{n} \frac{f_i\left(x_1^{(0)}, \ldots, x_{j-1}^{(0)}, y_j, \ldots, y_n\right) - f_i\left(x_1^{(0)}, \ldots, x_j^{(0)}, y_{j+1}, \ldots, y_n\right)}{y_j - x_j^{(0)}} \cdot \left(y_j - x_j^{(0)}\right)$$

the matrix $H(x^{(0)}, y)$ with $H(x^{(0)}, y)_{ij} :=$

$$\begin{cases} \dfrac{f_i\left(x_1^{(0)}, \ldots, x_{j-1}^{(0)}, y_j, \ldots, y_n\right) - f_i\left(x_1^{(0)}, \ldots, x_j^{(0)}, y_{j+1}, \ldots, y_n\right)}{y_j - x_j^{(0)}} & , y_j \neq x_j^{(0)} \\ c_{ij} & , y_j = x_j^{(0)} \end{cases}$$

fulfills $f(y) - f(x^{(0)}) = H(x^{(0)}, y) \cdot (y - x^{(0)})$ for any choice of $c_{ij} \in \mathbb{R}$. If $f'(x^{(0)})$ exists, it is natural to set $c_{ij} := \frac{\partial f_i}{\partial x_j}\left(x_1^{(0)}, \ldots, x_j^{(0)}, y_{j+1}, \ldots, y_n\right)$. If there exists an interval extension $H(x^{(0)}, [x])$ as described in Section 6.1, then the Krawczyk operator can be defined as

$$K(x^{(0)}, R, [x]) = x^{(0)} - R \cdot f(x^{(0)}) + (I - R \cdot H(x^{(0)}, [x])) \cdot ([x] - x^{(0)}).$$

The difference between $f'([x])$ and $H(x^{(0)}, [x])$ will be presented in the following example.

Example 6.7. Let $f(x) = x^2$, $x \in [x] \subseteq \mathbb{R}$. Then,

$$f(y) - f(x^{(0)}) = y^2 - (x^{(0)})^2 = (y + x^{(0)}) \cdot (y - x^{(0)})$$

and Hansen's method yields

$$H(x^{(0)}, [x]) = [x] + x^{(0)}.$$

Considering simple differentiation one gets

$$f'(x) = 2 \cdot x \quad \text{and} \quad f'([x]) = 2 \cdot [x].$$

For example, if $[x] = [3, 5]$ and $x^{(0)} = 4$, then Hansen's method yields

$$H(x^{(0)}, [x]) = [3, 5] + 4 = [7, 9],$$

whereas simple differentiation leads to

$$f'([x]) = 2 \cdot [3, 5] = [6, 10],$$

a result less precise than that of Hansen's result. □

6.3 The Moore Test

Let $f : D \to \mathbb{R}^n$, $D \subseteq \mathbb{R}^n$ be continuous. In 1977, R. Moore published a method for testing/proving the guess, if within a given interval vector $[x] \in \mathbf{IR}^n$, $[x] \subseteq D$, there exists a zero of f. It is based on the Krawczyk operator and on Brouwer's fixed point theorem.

Assume that we know an interval (slope) matrix $Y(x^0, [x])$, which fulfills the following: let $x^{(0)} \in [x] \subseteq D$ be fixed. Then, for any $y \in [x]$ there exists a matrix $Y(x^{(0)}, y) \in Y(x^{(0)}, [x])$ so that

$$f(y) - f(x^{(0)}) = Y(x^{(0)}, y) \cdot (y - x^{(0)}). \tag{6.10}$$

In Section 6.2 we have presented two possibilities for finding such an interval matrix. Independently of the choice of $Y(x^{(0)}, [x])$, the Krawczyk operator is defined as

$$K(\hat{x}, R, [x]) = \hat{x} - R \cdot f(\hat{x}) + (I - R \cdot Y(\hat{x}, [x])) \cdot ([x] - \hat{x})$$

where I denotes the unit matrix, $R \in \mathbb{R}^{n \times n}$ is a nonsingular matrix, and where $\hat{x} \in [x]$ is arbitrary, but fixed.

Theorem 6.1. (Moore, 1977) *If*

$$K(\hat{x}, R, [x]) \subseteq [x], \tag{6.11}$$

then there exists some $x^* \in [x]$ *with* $f(x^*) = 0$.

Proof: Multiplying (6.10) with R, we get after some calculations

$$-R \cdot f(y) = -R \cdot f(\hat{x}) - R \cdot Y(\hat{x}, y) \cdot (y - \hat{x}),$$

whence

$$y - R \cdot f(y) = \hat{x} - R \cdot f(\hat{x}) + (I - R \cdot Y(\hat{x}, y)) \cdot (y - \hat{x}).$$

Considering the continuous function

$$r(y) := y - R \cdot f(y), \quad y \in \mathbb{R}^n,$$

for any $y \in [x]$ we have

$$\begin{aligned} r(y) &= \hat{x} - R \cdot f(\hat{x}) + (I - R \cdot Y(\hat{x}, y)) \cdot (y - \hat{x}) \\ &\in \hat{x} - R \cdot f(\hat{x}) + (I - R \cdot Y(\hat{x}, [x])) \cdot ([x] - \hat{x}) \\ &= K(\hat{x}, R, [x]) \subseteq [x]. \end{aligned}$$

According to Brouwer's fixed point theorem, there exists some $x^* \in [x]$ with $r(x^*) = x^*$. Since R is nonsingular, we have $f(x^*) = 0$. □

The check for the validity of (6.11) is called Moore test. A way to find an interval vector $[x]$, for which it is likely to verify (6.11), has been published by G. Alefeld, A. Gienger, and F. Potra in 1994.

6.4 The Moore-Kioustelidis Test

If Brouwer's fixed point theorem can be used for verification methods, then so, too, can the Poincaré-Miranda theorem. This was remarked by J. B. Kioustelidis in 1978. Basically the idea is as follows.

Let $[x] \in \mathbf{IR}^n$, $f : [x] \to \mathbb{R}^n$ be continuous, and let the opposite faces $[x]^{i,\pm}$ be defined by

$$\left.\begin{aligned} [x]^{i,+} &:= \Big([x_1], \cdots, [x_{i-1}], \overline{x}_i, [x_{i+1}], \cdots, [x_n]\Big)^{\mathrm{T}}, \\ [x]^{i,-} &:= \Big([x_1], \cdots, [x_{i-1}], \underline{x}_i, [x_{i+1}], \cdots, [x_n]\Big)^{\mathrm{T}}, \end{aligned}\right\} \quad i = 1, ..., n. \tag{6.12}$$

Applying the Poincaré-Miranda theorem, one could try to verify for all $i = 1, ..., n$

$$\sup f_i([x]^{i,+}) \le 0 \le \inf f_i([x]^{i,-}) \tag{6.13}$$

or

$$\sup f_i([x]^{i,-}) \le 0 \le \inf f_i([x]^{i,+}), \tag{6.14}$$

using the interval extensions explained in Section 6.1. But it is very unlikely that this works as we will see in the following example.

Example 6.8. Let $f : \mathbb{R}^2 \to \mathbb{R}^2$ be defined as

$$f(x,y) = \begin{pmatrix} f_1(x,y) \\ f_2(x,y) \end{pmatrix} = \begin{pmatrix} x \cdot y \\ x + y - 1 \end{pmatrix}.$$

A zero is given by $(x^*, y^*)^{\mathrm{T}} = (1,0)^{\mathrm{T}}$. However, for any interval vector $[x] \in \mathbf{IR}^2$ with $(x^*, y^*)^{\mathrm{T}}$ as its center, neither (6.13) nor (6.14) can be valid. To show this, let

$$[x] = \begin{pmatrix} [1-\varepsilon, 1+\varepsilon] \\ [-\delta, \delta] \end{pmatrix} \quad \text{with arbitrarily small } \varepsilon > 0,\ \delta > 0.$$

Then, for any

$$\begin{pmatrix} x \\ y \end{pmatrix} \in [x]^{1,+} = \begin{pmatrix} 1+\varepsilon \\ [-\delta, \delta] \end{pmatrix}$$

we get

$$f_1(x,y) = f_1(1+\varepsilon, y) = (1+\varepsilon) \cdot y, \quad y \in [-\delta, \delta].$$

On the other hand, for any

$$\begin{pmatrix} x \\ y \end{pmatrix} \in [x]^{1,-} = \begin{pmatrix} 1-\varepsilon \\ [-\delta, \delta] \end{pmatrix}$$

we get

$$f_1(x,y) = f_1(1-\varepsilon, y) = (1-\varepsilon) \cdot y, \quad y \in [-\delta, \delta].$$

Thus,

$$f_1(1+\varepsilon, \frac{\delta}{2}) \cdot f_1(1-\varepsilon, \frac{\delta}{2}) > 0.$$

Therefore, the Poincaré-Miranda theorem cannot be applied. □

The idea of Kioustelidis was to apply the Poincaré-Miranda theorem not to f, but to some auxiliary function g. If f is differentiable, then

$$f(x) \approx f(x^c) + f'(x^c) \cdot (x - x^c).$$

Furthermore, if $f(x^c) \approx 0$ and if $f'(x^c)$ is nonsingular, then

$$x - x^c \approx \left(f'(x^c)\right)^{-1} \cdot f(x).$$

Setting $g(x) := \left(f'(x^c)\right)^{-1} \cdot f(x)$ we get

$$\left.\begin{array}{lll} g_i(x) \approx (x - x^c)_i = \underline{x}_i - x_i^c \leq 0 & \text{if} & x \in [x]^{i,-} \\ g_i(y) \approx (y - x^c)_i = \overline{x}_i - x_i^c \geq 0 & \text{if} & y \in [x]^{i,+} \end{array}\right\} \quad i = 1, ..., n.$$

This is exactly the behaviour that we need in order to apply one of the corollaries of the Poincaré-Miranda theorem.

Using a variation of this idea, R. Moore and J. B. Kioustelidis established the following.

Theorem 6.2. (Moore and Kioustelidis, 1980) *Let $\hat{x}, s, t \in \mathbb{R}^n$, $s \geq 0$, $t \geq 0$, and $[x] = [\hat{x} - s, \hat{x} + t] \in \mathbf{IR}^n$. For $i = 1, ..., n$ the interval vectors $[x]^{i,+}$ and $[x]^{i,-}$ are defined as in* (6.12). *Let $f : D \to \mathbb{R}^n$, $D \subseteq \mathbb{R}^n$ be continuous with $[x] \subseteq D$. Assume, that an interval (slope) matrix $Y(x^0, [x])$ is known, which fulfills the following. Let $x^{(0)} \in [x]$ be fixed. Then, for any $y \in [x]$ there exists a matrix $Y(x^{(0)}, y) \in Y(x^{(0)}, [x])$ so that (6.10) holds. Let e_i denote the ith unit vector, $R \in \mathbb{R}^{n \times n}$ be nonsingular, $g(x) := R \cdot f(x)$, and*

$$\begin{aligned} [l]^{i,+} &:= g_i(\hat{x} + t_i e_i) + \sum_{j=1,\, j \neq i}^{n} \left(R \cdot Y(\hat{x} + t_i e_i, [x]^{i,+})\right)_{ij} \cdot [-s_j, t_j], \\ [l]^{i,-} &:= g_i(\hat{x} - s_i e_i) + \sum_{j=1,\, j \neq i}^{n} \left(R \cdot Y(\hat{x} - s_i e_i, [x]^{i,-})\right)_{ij} \cdot [-s_j, t_j]. \end{aligned}$$

If for $i = 1, ..., n$ it holds that

$$\sup[l]^{i,-} \leq 0 \leq \inf[l]^{i,+} \tag{6.15}$$

or

$$\sup[l]^{i,+} \leq 0 \leq \inf[l]^{i,-}, \tag{6.16}$$

then there exists some $x^* \in [x]$ with $f(x^*) = 0$.

Proof: Let $i \in \{1, ..., n\}$ be arbitrary, but fixed. Setting $x^{(0)} = \hat{x} + t_i e_i$ in (6.10), then for every $x \in [x]^{i,+}$ it holds

$$f(x) = f(\hat{x} + t_i e_i) + Y(\hat{x} + t_i e_i, x) \cdot (x - \hat{x} - t_i e_i).$$

After multiplying by R, we get

$$\begin{aligned} g(x) &= g(\hat{x} + t_i e_i) + (R \cdot Y(\hat{x} + t_i e_i, x)) \cdot (x - \hat{x} - t_i e_i) \\ &\in g(\hat{x} + t_i e_i) + (R \cdot Y(\hat{x} + t_i e_i, [x]^{i,+})) \cdot ([x]^{i,+} - \hat{x} - t_i e_i). \end{aligned}$$

Since

$$[x]^{i,+} - \hat{x} - t_i e_i = \begin{pmatrix} [-s_1, t_1] \\ \vdots \\ [-s_{i-1}, t_{i-1}] \\ 0 \\ [-s_{i+1}, t_{i+1}] \\ \vdots \\ [-s_n, t_n] \end{pmatrix},$$

it follows that $g_i(x) \in [l]^{i,+}$ for all $x \in [x]^{i,+}$. Analogously, we get $g_i(y) \in [l]^{i,-}$ for all $y \in [x]^{i,-}$. But (6.15) and (6.16) assure us that

$$g_i(x) \cdot g_i(y) \leq 0 \quad \text{for all } x \in [x]^{i,+} \quad \text{and for all } y \in [x]^{i,-}.$$

By Corollary 4.2 it follows that there exists some $x^* \in [x]$ with $g(x^*) = 0$. Since R is nonsingular, $f(x^*) = 0$ as well. □

The check for validity of (6.15) and (6.16) is called Moore-Kioustelidis test. A comparison to the Moore test will be given in Section 6.6. Beforehand we present another verification test also based on the Poincaré-Miranda theorem.

6.5 The Frommer-Lang-Schnurr Test

Theorem 6.3. (Frommer, Lang, and Schnurr, 2004) *Let $[x] \in \mathbf{IR}^n$ and for $i = 1, ..., n$ let the interval vectors $[x]^{i,+}$ and $[x]^{i,-}$ be defined as in (6.12). Let $f : D \to \mathbb{R}^n$, $D \subseteq \mathbb{R}^n$ be continuous with $[x] \subseteq D$. Assume, that an interval (slope) matrix $Y(x^{(0)}, [x])$ satisfies: Let $x^{(0)} \in [x]$ be fixed. Then, for any $y \in [x]$ there exists a matrix $Y(x^{(0)}, y) \in Y(x^{(0)}, [x])$ so that (6.10) holds. Let $\hat{x} \in [x]$, $R \in \mathbb{R}^{n\times n}$ be nonsingular, $g(x) := R \cdot f(x)$, and*

$$\begin{aligned} [m(\hat{x})]^{i,+} &:= g_i(\hat{x}) + \sum_{j=1}^{n} \Big(R \cdot Y(\hat{x}, [x])\Big)_{ij} \cdot ([x_j]^{i,+} - \hat{x}_j), \\ [m(\hat{x})]^{i,-} &:= g_i(\hat{x}) + \sum_{j=1}^{n} \Big(R \cdot Y(\hat{x}, [x])\Big)_{ij} \cdot ([x_j]^{i,-} - \hat{x}_j). \end{aligned}$$

If for $i = 1, ..., n$

$$\sup[m(\hat{x})]^{i,-} \leq 0 \leq \inf[m(\hat{x})]^{i,+} \tag{6.17}$$

or

$$\sup[m(\hat{x})]^{i,+} \leq 0 \leq \inf[m(\hat{x})]^{i,-}, \tag{6.18}$$

then there exists some $x^* \in [x]$ with $f(x^*) = 0$.

Proof: Let $i \in \{1, ..., n\}$. According to (6.10) we get

$$g_i(x) \in [m(\hat{x})]^{i,+} \quad \text{for all } x \in [x]^{i,+}$$

and

$$g_i(y) \in [m(\hat{x})]^{i,-} \quad \text{for all } y \in [x]^{i,-}.$$

Calling on Corollary 4.2, (6.17), and (6.18) we conclude that there exists some $x^* \in [x]$ with $g(x^*) = 0$. Since R is nonsingular, it follows that $f(x^*) = 0$. □

The check for the validity of (6.17) and (6.18) is called the Frommer-Lang-Schnurr test.

6.6 A Comparison of the Tests

Theorem 6.4. *The Frommer-Lang-Schnurr test is more powerful than the Moore test, meaning that (6.11) implies (6.17).*

Proof: Suppose that the Moore test is successful. Then, considering the ith component one obtains

$$\underline{x}_i \le \inf K(\hat{x}, R, [x])_i = \inf \Big(\hat{x} - R \cdot f(\hat{x}) + (I - R \cdot Y(\hat{x}, [x])) \cdot ([x] - \hat{x})\Big)_i .$$

Since $g(x) = R \cdot f(x)$ we have

$$K(\hat{x}, R, [x])_i = \hat{x}_i - g_i(\hat{x}) + \sum_{j=1}^{n} (I - R \cdot Y(\hat{x}, [x]))_{ij} \cdot ([x_j] - \hat{x}_j),$$

and since for $j \neq i$ we have $[x_j]^{i,-} = [x_j]$, we can conclude that $K(\hat{x}, R, [x])_i =$

$$\hat{x}_i - g_i(\hat{x}) + (I - R \cdot Y(\hat{x}, [x]))_{ii} \cdot ([x_i] - \hat{x}_i) - \sum_{j=1, j\neq i}^{n} (R \cdot Y(\hat{x}, [x]))_{ij} \cdot ([x_j]^{i,-} - \hat{x}_j).$$

Furthermore, due to

$$(I - R \cdot Y(\hat{x}, [x]))_{ii} \cdot ([x_i] - \hat{x}_i) \supseteq (I - R \cdot Y(\hat{x}, [x]))_{ii} \cdot (\underline{x}_i - \hat{x}_i)$$

we have

$\inf K(\hat{x}, R, [x])_i \le$

$$\inf \Big(\hat{x}_i - g_i(\hat{x}) + (I - R \cdot Y(\hat{x}, [x]))_{ii} \cdot (\underline{x}_i - \hat{x}_i) - \sum_{j=1, j\neq i}^{n} (R \cdot Y(\hat{x}, [x]))_{ij} \cdot ([x_j]^{i,-} - \hat{x}_j)\Big).$$

From the distributive law $[a] \cdot \gamma + [b] \cdot \gamma = ([a] + [b]) \cdot \gamma$ holding for the scalar $\gamma = \underline{x}_i - \hat{x}_i$ and the fact that $\underline{x}_i = [x_i]^{i,-}$ it follows that

$$\underline{x}_i \leq \inf K(\hat{x}, R, [x])_i \leq \inf \left(\underline{x}_i - [m(\hat{x})]^{i,-}\right) = \underline{x}_i - \sup[m(\hat{x})]^{i,-}.$$

Thus,

$$\sup[m(\hat{x})]^{i,-} \leq 0.$$

In the same way, one can show that

$$\sup K(\hat{x}, R, [x])_i \leq \overline{x}_i$$

implies $0 \leq \inf[m(\hat{x})]^{i,+}$. □

The Moore test and the Frommer-Lang-Schnurr test require roughly the same computational work. Therefore, the Frommer-Lang-Schnurr test is more powerful than the Moore test by Theorem 6.4.

Comparing the Moore-Kioustelidis test with the Frommer-Lang-Schnurr test, it is obvious that the latter requires less computational work, since there the function $g(\cdot)$ and the interval (slope) matrix $Y(\cdot,\cdot)$ are only evaluated once.

However, the following theorem will present a special situation, where the Moore-Kioustelidis test is more powerful than the Frommer-Lang-Schnurr test.

Theorem 6.5. *Let $f : D \to \mathbb{R}^n$, $D \subseteq \mathbb{R}^n$ be differentiable and let $[x] \subseteq D$. Furthermore, let*

$$Y(\hat{x}, [x]) = f'([x]).$$

Then, concerning Theorems 6.3 and 6.2, we have

$$[l]^{i,+} \subseteq [m(\hat{x})]^{i,+} \quad \text{and} \quad [l]^{i,-} \subseteq [m(\hat{x})]^{i,-}. \tag{6.19}$$

This means, if (6.17) from Theorem 6.3 is valid, then the condition (6.15) from Theorem 6.2 is fulfilled; analogously, if (6.18) from Theorem 6.3 is valid, then the condition (6.16) from Theorem 6.2 is fulfilled.

Proof: Let $i \in \{1, ..., n\}$. According to (6.10) we have

$$f(\hat{x} + t_i e_i) - f(\hat{x}) = Y(\hat{x}, \hat{x} + t_i e_i) \cdot (\hat{x} + t_i e_i - \hat{x}).$$

After multiplying by R, we get

$$g(\hat{x} + t_i e_i) = g(\hat{x}) + (R \cdot Y(\hat{x}, \hat{x} + t_i e_i)) \cdot t_i e_i,$$

whence

$$g_i(\hat{x} + t_i e_i) \in g_i(\hat{x}) + t_i \cdot \left(R \cdot Y(\hat{x}, [x])\right)_{ii} = g_i(\hat{x}) + t_i \cdot \left(R \cdot f'([x])\right)_{ii}. \tag{6.20}$$

Since, in general, we have for $[a]$, $[b] \in \mathbf{IR}$ and any $\gamma \in \mathbb{R}$, that $\gamma \in [a]$ implies $\gamma + [b] \subseteq [a] + [b]$, we get from (6.20)

$$g_i(\hat{x} + t_i e_i) + \sum_{j=1,\, j \neq i}^{n} \left(R \cdot f'([x])\right)_{ij} \cdot [-s_j, t_j] \subseteq [m(\hat{x})]^{i,+}. \tag{6.21}$$

Using $f'([x]^{i,+}) \subseteq f'([x])$ we finally get $[l]^{i,+} \subseteq [m(\hat{x})]^{i,+}$. Analogously, one can show that $[l]^{i,-} \subseteq [m(\hat{x})]^{i,-}$. □

The following figure gives a summary for the three verification methods presented, where $\boxed{\text{A}} \Rightarrow \boxed{\text{B}}$ means that the success of test A implies the success of test B.

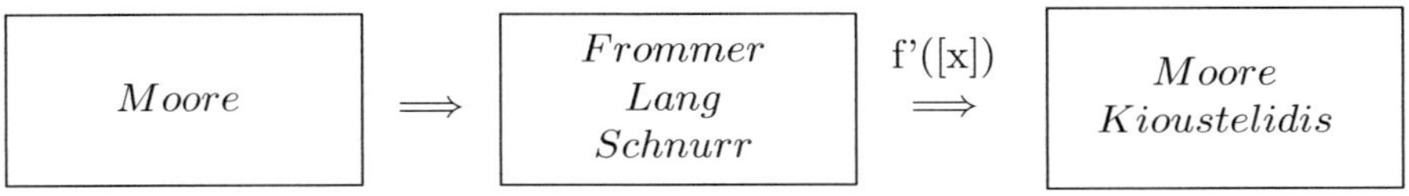

The following example will present a counterexample concerning the second implication when Hansen's slope matrix $H(\hat{x}, [x])$ is used as $Y(\hat{x}, [x])$ instead of $f'([x])$.

Example 6.9. (Schnurr, 2005) We consider the function

$$f(x_1, x_2) = \begin{pmatrix} f_1(x_1, x_2) \\ f_2(x_1, x_2) \end{pmatrix} = \begin{pmatrix} (4x_1 + 0.1)x_2^2 - (x_1 + 0.025)x_1^2 + 0.0525x_1 \\ x_2 + 0.01 \end{pmatrix}$$

and a point $\hat{x} = (\hat{x}_1\, \hat{x}_2)^{\mathrm{T}}$. The Jacobian matrix is

$$f'(x_1, x_2) = \begin{pmatrix} 4x_2^2 - 3x_1^2 - 0.05x_1 + 0.0525 & 2x_2\,(4x_1 + 0.1) \\ 0 & 1 \end{pmatrix},$$

whereas the slope matrix $H(\hat{x}, y)$ of Hansen's method is given by

$$\frac{f_1(y_1, y_2) - f_1(\hat{x}_1, y_2)}{y_1 - \hat{x}_1} =$$

$$4y_2^2 - y_1^2 - y_1\hat{x}_1 - \hat{x}_1^2 - 0.025 \cdot (y_1 + \hat{x}_1) + 0.0525,$$

$$\frac{f_1(\hat{x}_1, y_2) - f_1(\hat{x}_1, \hat{x}_2)}{y_2 - \hat{x}_2} = (4\hat{x}_1 + 0.1)\,(y_2 + \hat{x}_2),$$

$$\frac{f_2(y_1, y_2) - f_2(\hat{x}_1, y_2)}{y_1 - \hat{x}_1} = 0, \quad \frac{f_2(\hat{x}_1, y_2) - f_2(\hat{x}_1, \hat{x}_2)}{y_2 - \hat{x}_2} = 1.$$

Concerning $[x] = ([-0.1, 0.1]\,[-0.1, 0.1])^{\mathrm{T}}$ and $\hat{x} = 0$, we compare the Moore test with the Moore-Kioustelidis test. In both tests we use

$$R = \left(f^{'}(0)\right)^{-1} = \begin{pmatrix} \frac{1}{0.0525} & 0 \\ 0 & 1 \end{pmatrix}.$$

Then,

$$\frac{f_1(y_1, y_2) - f_1(0, y_2)}{y_1 - 0} = 4y_2^2 - (y_1 + 0.0125)^2 + 0.0525 + 0.0125^2.$$

Therefore, we have

$$Y(0, [x]) = \begin{pmatrix} [0.04\,, 0.0925 + 0.0125^2] & [-0.01, 0.01] \\ 0 & 1 \end{pmatrix}$$

and

$$\begin{aligned} I - R \cdot Y(0, [x]) &= \begin{pmatrix} \left[1 - \frac{0.0925+0.0125^2}{0.0525}\,, 1 - \frac{0.04}{0.0525}\right] & \left[-\frac{0.01}{0.0525}, \frac{0.01}{0.0525}\right] \\ 0 & 0 \end{pmatrix} \\ &\subseteq \begin{pmatrix} [-0.77\,, 0.24] & [-0.2, 0.2] \\ 0 & 0 \end{pmatrix}. \end{aligned}$$

Hence,

$$\begin{aligned} K(0, R, [x]) &= 0 - R \cdot f(0) + (I - R \cdot Y(0, [x])) \cdot \begin{pmatrix} [-0.1, 0.1] \\ [-0.1, 0.1] \end{pmatrix} \\ &\subseteq \begin{pmatrix} [-0.077, 0.077] + [-0.02, 0.02] \\ -0.01 \end{pmatrix} \subseteq [x]. \end{aligned}$$

Therefore, the Moore test is successful, and due to Theorem 6.4 also the Frommer-Lang-Schnurr test is successsful to verify that a zero x^* of f is within $[x]$. On the other hand, Theorem 6.2 says that

$$[l]^{1,+} = \left(R \cdot f(0.1, 0)\right)_1 + \left(R \cdot Y(\begin{pmatrix} 0.1 \\ 0 \end{pmatrix}, \begin{pmatrix} 0.1 \\ [-0.1, 0.1] \end{pmatrix})\right)_{12} \cdot [-0.1, 0.1]$$

$$= \left(\frac{1}{0.0525} \cdot 4 \cdot 10^{-3}\right) + \left(\frac{1}{0.0525} \cdot 0.5 \cdot [-0.1, 0.1]\right) \cdot [-0.1, 0.1] = \frac{1}{105}[-2, 18].$$

Thus, the Moore-Kioustelidis test fails. □

Remark 6.1. Being aware of the fact that Brouwer's fixed point theorem as well as the Poincaré-Miranda theorem can be proved (very briefly) by using the so-called degree of a mapping (see the book of J. Cronin and the paper of M. N. Vrahatis, respectively), one might get the idea to verify the existence of a zero directly by using the degree of a mapping and interval arithmetic. This was done by A. Frommer, F. Hoxha, and B. Lang.

6.7 Final Remarks

To see the full power of interval computations, one has to use a computer. There are several programming languages, that support interval arithmetic. For example, PASCAL-XSC, C-XSC, and INTLAB. See, for example, the book of R. Hammer, M. Hocks, U. Kulisch, and D. Ratz, the book of R. Klatte, U. Kulisch, A. Wiethoff, C. Lawo, and M. Rauch, and the paper of S. Rump, respectively.

These programming languages are extensions of PASCAL, C, and Matlab, respectively, by introducing another data type called interval. If $[a]$, $[b]$, and $[c]$ are declared as an interval, then the operations (6.2) are predefined and the boundaries of $[c] = [a] * [b]$ with $* \in \{+, -, \cdot, /\}$ are automatically rounded in the corresponding direction, for the case, that $\underline{c}$ and $\overline{c}$ have to be rounded. For example, $\underline{a} + \underline{b}$ is rounded downwardly and $\overline{a} + \overline{b}$ is rounded upwardly to the next machine number. This means, if $\underline{a}$, $\overline{a}$, $\underline{b}$, and $\overline{b}$ are machine numbers, then $a + b \in [a] + [b]$ for all real $a \in [\underline{a}, \overline{a}]$ and all real $b \in [\underline{b}, \overline{b}]$.

Using interval arithmetic by a computer, one has to extend the ideas of the preceding sections a little bit.

Example 6.10. Consider the function

$$g(x) = 3x - \sqrt{2}, \quad x \in \mathbb{R}.$$

We want to verify, that within the interval $[x] = [0.47, 0.48]$ there is a zero of g. Since within a computer the number $\sqrt{2}$ cannot be represented, the test

$$g(0.47) \cdot g(0.48) \leq 0$$

cannot be done exactly by a computer. Therefore, the number $\sqrt{2}$ is included in an interval; i.e.,

$$\sqrt{2} \in [1.4142, 1.4143],$$

and the function is extended to the function $g(x; a)$ with $a \in [1.4142, 1.4143]$, $x \in \mathbb{R}$. If it is possible to show that for all $a \in [1.4142, 1.4143]$ it holds that

$$g(0.47; a) \cdot g(0.48; a) \leq 0,$$

then one can conclude that for all $a \in [1.4142, 1.4143]$ there is a zero in $[x]$; especially for $a = \sqrt{2}$. □

This means, actually we have to consider functions $f(x_1, ..., x_n; a_1, ..., a_m)$ with $(x_1, ..., x_n)^{\mathrm{T}} \in D$, where D is the domain of the function and where a_1,...,a_m are parameters, that can vary in given intervals. Indeed, it is easy to extend the corresponding theorems to this generalized case. However, the hierarchy presented in the preceding section is no longer true as we will see in the following example.

Example 6.11. We consider the continuous function

$$f(x;a) = \left\{ \begin{array}{ll} 2x-1, & x \geq 1 \\ a \cdot x - a + 1, & x < 1 \end{array} \right\} \quad a \in [a] = [2,4].$$

Taking

$$f(x;[a]) = \left\{ \begin{array}{ll} 2x-1, & x \geq 1 \\ [a] \cdot x - [a] + 1, & x < 1 \end{array} \right.$$

and

$$Y([x];[a]) = \text{convex hull of } \{2,[a]\} = [2,4],$$

we want to verify that within the interval $[x] = [\frac{1}{2}, 2]$ there is a zero of f. Choosing $R \neq 0$ and $\hat{x} = 1$ we have by Theorem 6.3

$$[m(1)]^{+} = R \cdot 1 + R \cdot [2,4] \cdot (2-1) = R \cdot [3,5]$$

and

$$[m(1)]^{-} = R \cdot 1 + R \cdot [2,4] \cdot (\frac{1}{2} - 1) = R \cdot [-1,0],$$

whereas by Theorem 6.2 we have

$$[l]^{-} = R \cdot ([a] \cdot \frac{1}{2} - [a] + 1) = R \cdot [-2,1].$$

This means, that the Frommer-Lang-Schnurr test is successful, whereas the Moore-Kioustelidis test fails. □

For a hierachy in the general case we refer to (82). See also (80) for some numerical examples.

For more examples where interval arithmetic is used to prove conjectures via the computer we refer to (37).

7 Banach's Fixed Point Theorem

Using Brouwer's fixed point theorem one can attack the problem of showing the existence of a fixed point. As already mentioned in Chapter 2 a fixed point verified by Brouwer's theorem is not necessarily unique. See Figure 2.2. The question naturally arises how the assumptions can be strengthened so that a verified fixed point is unique.

To get the idea, let x^* be a fixed point of $f : [a, b] \to [a, b]$. In order to guarantee that x^* is the only fixed point, the problem $f(x) = x$ must not have a second solution. If there was a second solution, say $\tilde{x}$, $\tilde{x} \neq x^*$ it would follow that

$$1 = \frac{\tilde{x} - x^*}{\tilde{x} - x^*} = \frac{f(\tilde{x}) - f(x^*)}{\tilde{x} - x^*}.$$

If we assume f to be differentiable, the mean-value theorem says that there exists ξ between $\tilde{x}$ and x^* so that

$$\frac{f(\tilde{x}) - f(x^*)}{\tilde{x} - x^*} = f'(\xi).$$

So, if $f : [a, b] \to [a, b]$ is differentiable satisfying

$$f'(x) < 1 \quad \text{for all } x \in [a, b], \tag{7.1}$$

then f has exactly one fixed point in $[a, b]$.

How can this idea be extended to the n-dimensional case?

Since, for $\tilde{x} > x^*$,

$$\frac{f(\tilde{x}) - f(x^*)}{\tilde{x} - x^*} = f'(\xi) < 1$$

is equivalent to

$$f(\tilde{x}) - f(x^*) = f'(\xi) \cdot (\tilde{x} - x^*) < (\tilde{x} - x^*),$$

a (more or less obvious) generalization to the n-dimensional case is

$$\|f(\tilde{x}) - f(x^*)\| < \|\tilde{x} - x^*\| \quad \text{if } \tilde{x} \neq x^* \tag{7.2}$$

for some norm in $\mathbb{R}^n$. (Note, however, that in (7.1) it reads $f'(x) < 1$ and not $|f'(x)| < 1$.) So, trivially, we have the following theorem.

Theorem 7.1. *Let $K \subseteq \mathbb{R}^n$ be nonenpty, convex, bounded, and closed. Furthermore, let $f : K \to K$ satisfy*

$$\|f(x) - f(y)\| < \|x - y\| \tag{7.3}$$

for all $x, y \in K$, $x \neq y$ for some norm in $\mathbb{R}^n$. Then, f has exactly one fixed point in K.

Proof: By (7.3) the function f is continuous. Therefore, Brouwer's fixed point theorem guarantees the existence of a fixed point, say x^*. If there was another fixed point, say $\tilde{x}$, with $\tilde{x} \neq x^*$, then it would follow due to (7.3)

$$\|\tilde{x} - x^*\| = \|f(\tilde{x}) - f(x^*)\| < \|\tilde{x} - x^*\|$$

which is a contradiction. □

Of course, if the proof is so simple, there is no reason to name this result after a mathematician. So, there must be more!

Indeed, the astonishing thing is (beyond the fact that the fixed point can be approximated iteratively, see Theorem 7.2) the fact that the result is also true in some vector spaces that can be infinite-dimensional. These vector spaces are called Banach spaces. We will introduce them in the following section. In order to emphasize this result, we mention that Brouwer's fixed point theorem is not valid in infinite-dimensional vector spaces.

7.1 Banach Spaces

Before we introduce some Banach spaces, we briefly remember the definition of vector spaces and normed spaces. Already in school $\mathbb{R}^2$ was considered as a vector space. For example,

$$\begin{pmatrix} 4 \\ 2 \end{pmatrix} \in \mathbb{R}^2.$$

Then, $2 \cdot \binom{4}{2}$ points in the same direction but with doubled length. Also the vector addition is well known from school. For example

$$\begin{pmatrix} 4 \\ 2 \end{pmatrix} + \begin{pmatrix} 2 \\ 6 \end{pmatrix} = \begin{pmatrix} 6 \\ 8 \end{pmatrix} \in \mathbb{R}^2.$$

See also Figure 7.1.

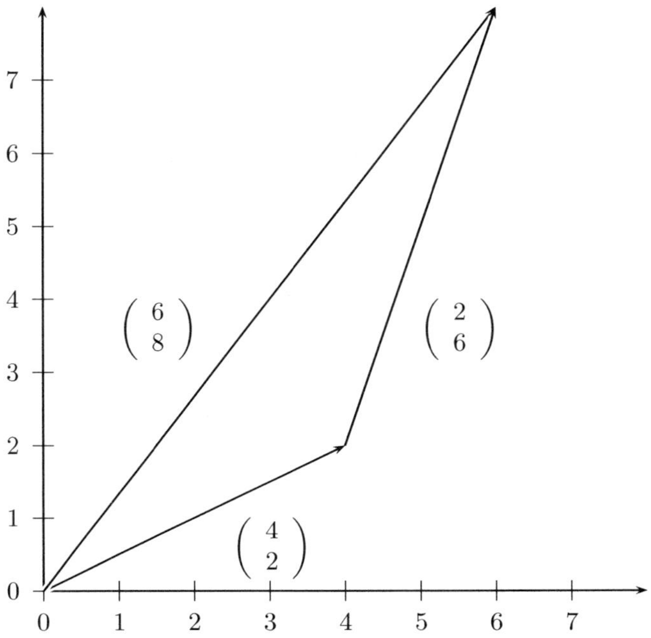

Figure 7.1: **Vector addition visualized in $\mathbb{R}^2$**

Let $E = \mathbb{R}^2$, the basic observation for a generalization is the fact that

$$\begin{aligned} x \in E,\, y \in E \quad &\Rightarrow \quad x + y \in E, \\ \lambda \in \mathbb{R},\, x \in E \quad &\Rightarrow \quad \lambda \cdot x \in E. \end{aligned} \tag{7.4}$$

Trivially, (7.4) can be verified for $E = \mathbb{R}^n$ for all integers $n \geq 2$. More important, however, is the fact, that we can also consider the set of all real-valued continuous functions with domain $[a, b]$, denoted by $C([a, b])$, as E. If $f,\, g \in C([a, b])$ and $\lambda \in \mathbb{R}$, then also

$$f + g \in C([a, b]) \quad \text{and} \quad \lambda \cdot f \in C([a, b])$$

by defining

$$(f + g)(x) := f(x) + g(x) \quad \text{for all } x \in [a, b]$$

and

$$(\lambda \cdot f)(x) := \lambda \cdot f(x) \quad \text{for all } x \in [a, b].$$

Satisfying (7.4) is not the only property that the sets $\mathbb{R}^n$ and $C([a, b])$ have in common. The idea is to find as many common properties as possible and to prove theorems only based on these properties. Formally, a real vector space (sometimes also called a real linear space) is a nonempty set E fulfilling (7.4) and

$$x + (y + z) = (x + y) + z \text{ for all } x,\ y,\ z \in E,$$
$$x + y = y + x \text{ for all } x,\ y \in E,$$
$$\text{there exists } 0 \in E \text{ so that } x + 0 = x \text{ for all } x \in E,$$
$$\text{for all } x \in E \text{ there exists } -x \in E \text{ so that } x + (-x) = 0,$$
$$\lambda \cdot (x + y) = \lambda \cdot x + \lambda \cdot y \text{ for all } x,\ y \in E \text{ and all } \lambda \in \mathbb{R},$$
$$(\lambda + \nu) \cdot x = \lambda \cdot x + \nu \cdot x \text{ for all } x \in E \text{ and for all } \lambda,\ \nu \in \mathbb{R},$$
$$(\lambda \cdot \nu) \cdot x = \lambda \cdot (\nu \cdot x) \text{ for all } x \in E \text{ and for all } \lambda,\ \nu \in \mathbb{R},$$
$$1 \cdot x = x \text{ for all } x \in E.$$

A real vector space E is called a normed vector space, if there is a function $\|\cdot\| : E \to \mathbb{R}$ satisfying

$$\begin{aligned} \|x\| &\geq 0 \text{ and } \|x\| = 0 \Leftrightarrow x = 0, \\ \|\lambda \cdot x\| &= |\lambda| \cdot \|x\| \\ \|x + y\| &\leq \|x\| + \|y\|. \end{aligned}$$

Now, we can give the definition of a Banach space.

Definition 7.1. *A normed vector space $(E, \|\cdot\|)$ is called a Banach space, if any so-called Cauchy sequence in E (see below) is convergent to a limit that belongs to E; where a sequence $\{x^{(n)}\}_{n=1}^{\infty}$ is called a Cauchy sequence if the following holds: For any $\varepsilon > 0$ there exists $n_0(\varepsilon)$ so that*

$$\|x^{(m)} - x^{(n)}\| < \varepsilon \quad \text{if } m,\ n \geq n_0(\varepsilon).$$

This definition is enough to extend Theorem 7.1.

Theorem 7.2. (Banach's fixed point theorem) *Let $(E, \|\cdot\|)$ be a Banach space, and let $D \subseteq E$ be nonempty and closed. If the function $T : D \to D$ satisfies*

$$\|T(x) - T(y)\| \leq q \cdot \|x - y\| \quad \text{for all } x, y \in D \tag{7.5}$$

with some $q < 1$, then within D there exists a unique fixed point x^* of T. Furthermore, for any $x^{(0)} \in D$ the so-called fixed point iteration

$$x^{(n+1)} := T(x^{(n)}), \quad n = 0, 1, 2, ...$$

converges to x^* and

$$\|x^{(n)} - x^*\| \leq \frac{1}{1-q} \cdot \|x^{(n+1)} - x^{(n)}\| \leq \frac{q^n}{1-q} \cdot \|x^{(1)} - x^{(0)}\|.$$

Proof: First, we note that from (7.5) it follows that T is continuous. Then, we choose arbitrarily $x^{(0)} \in D$ and consider the sequence

$$x^{(n+1)} := T(x^{(n)}), \quad n = 0, 1, 2, 3, \dots .$$

This sequence is well-defined, since T is a self-mapping. From

$$\|x^{(n+1)} - x^{(n)}\| = \|T(x^{(n)}) - T(x^{(n-1)})\| \leq q \cdot \|x^{(n)} - x^{(n-1)}\|$$

we get by induction

$$\|x^{(n+1)} - x^{(n)}\| \leq q^n \cdot \|x^{(1)} - x^{(0)}\|. \tag{7.6}$$

Now, for any $p > 0$ we have $\|x^{(n+p)} - x^{(n)}\| \leq$

$$\begin{aligned} & \|x^{(n+p)} - T(x^{(n+p)})\| + \|T(x^{(n+p)}) - T(x^{(n)})\| + \|T(x^{(n)}) - x^{(n)}\| \\ \leq \; & \|x^{(n+p)} - T(x^{(n+p)})\| + q \cdot \|x^{(n+p)} - x^{(n)}\| + \|T(x^{(n)}) - x^{(n)}\| \\ = \; & \|x^{(n+p)} - x^{(n+p+1)}\| + q \cdot \|x^{(n+p)} - x^{(n)}\| + \|x^{(n+1)} - x^{(n)}\|. \end{aligned}$$

So, by (7.6) it follows

$$\|x^{(n+p)} - x^{(n)}\| \leq \frac{1}{1-q}(q^{n+p} + q^n) \cdot \|x^{(1)} - x^{(0)}\| = \frac{q^p + 1}{1-q} \cdot \|x^{(1)} - x^{(0)}\| \cdot q^n \leq C \cdot q^n$$

with

$$C = \frac{2}{1-q} \cdot \|x^{(1)} - x^{(0)}\|.$$

Hence, $\{x^{(n)}\}_{n=0}^{\infty}$ is a Cauchy sequence. The latter has a limit, say x^*, since E is a Banach space, and $x^* \in D$, since D is closed. Due to the continuity of T we have

$$T(x^*) = T(\lim_{n \to \infty} x^{(n)}) = \lim_{n \to \infty} T(x^{(n)}) = \lim_{n \to \infty} x^{(n+1)} = x^*.$$

This means, that $x^* \in D$ is a fixed point of T.

To show uniqueness, we assume that T has two fixed points, say x^* and y^*. Then, with (7.5) it follows

$$\|x^* - y^*\| = \|T(x^*) - T(y^*)\| \leq q \cdot \|x^* - y^*\|.$$

However, this can only be true if $x^* = y^*$, since $q < 1$.

To get the related error estimate we consider

$$\begin{aligned} \|x^{(n)} - x^*\| & \leq \|x^{(n)} - x^{(n+1)}\| + \|x^{(n+1)} - x^*\| \\ & = \|x^{(n+1)} - x^{(n)}\| + \|T(x^{(n)}) - T(x^*)\| \\ & \leq \|x^{(n+1)} - x^{(n)}\| + q \cdot \|x^{(n)} - x^*\|. \end{aligned}$$

Hence,

$$\|x^{(n)} - x^*\| \leq \frac{1}{1-q} \cdot \|x^{(n+1)} - x^{(n)}\|,$$

and (7.6) finishes the proof. □

Banach's fixed point theorem gets its full power, if one is aware of some Banach spaces. We name just a few. (For a proof of the fact that these vector spaces are indeed Banach spaces, we refer to any book on functional analysis).

- $(\mathbb{R}^n, \|\cdot\|)$, where $\|\cdot\|$ can be any norm in $\mathbb{R}^n$.
- $(l^2, \|\cdot\|)$, where l^2 denotes the vector space of all square summable sequences of real numbers; i.e.,

  $$x = \{x_i\}_{i=1}^{\infty} \in l^2 :\Leftrightarrow \sum_{i=1}^{\infty} x_i^2 < \infty,$$

  and where

  $$\|x\| := \sqrt{\sum_{i=1}^{\infty} x_i^2}.$$

- $(l^\infty, \|\cdot\|)$, where l^∞ denotes the vector space of all bounded sequences of real numbers, and where

  $$\|x\| := \sup_{i=1}^{\infty} |x_i|.$$

- $(c_0, \|\cdot\|)$, where c_0 denotes the vector space of all real sequences converging to zero, and where

  $$\|x\| := \sup_{i=1}^{\infty} |x_i|.$$

- $(C([a,b]), \|\cdot\|_\infty)$, where $C([a,b])$ denotes the vector space of all real-valued, continuous functions with domain $[a,b]$, and where

  $$\|x\|_\infty := \max\{|x(t)| : t \in [a,b]\}.$$

7.2 Ordinary Differential Equations

The breakthrough of Banach's fixed point theorem is in the field of ordinary differential equations. This will be presented in the following example.

Example 7.1. Given $\xi, \eta, a \in \mathbb{R}$ and $f : [\xi, \xi + a] \times \mathbb{R} \to \mathbb{R}$. Then, the initial value problem is to find a function $y : [\xi, \xi + a] \to \mathbb{R}$ satisfying

$$\begin{aligned} y'(x) &= f(x, y(x)), \quad x \in [\xi, \xi + a] \\ y(\xi) &= \eta. \end{aligned}$$

A famous result by Picard-Lindelöf says that this problem has a unique solution, if $f(x, y)$ is continuous and if there exists $L \geq 0$ so that

$$|f(x, y) - f(x, \overline{y})| \leq L \cdot |y - \overline{y}|$$

is valid for all $x \in [\xi, \xi + a]$ and for all $y, \overline{y} \in \mathbb{R}$. Its proof can be done by using Banach's fixed point theorem as follows. We consider the Banach space $C([\xi, \xi + a])$ with the weighted norm

$$\|y\|_w := \max\{|y(x)| \cdot e^{-2L \cdot x}, \, x \in [\xi, \xi + a]\}.$$

Since f is continuous, $y(x)$ is a solution of the initial value problem, if and only if

$$y(x) = \eta + \int_{\xi}^{x} f(t, y(t))dt.$$

So, we define the function

$$T : C([\xi, \xi + a]) \to C([\xi, \xi + a])$$

by

$$T(y)(x) := \eta + \int_{\xi}^{x} f(t, y(t))dt.$$

Then, the initial value problem has a unique solution, if and only if T has a unique fixed point.

To verify the existence of a unique fixed point of T, let $y, z \in C([\xi, \xi+a])$. Then,

$$\begin{aligned}
|T(y)(x) - T(z)(x)| &= \left| \int_{\xi}^{x} \Big(f(t, y(t)) - f(t, z(t)) \Big) dt \right| \\
&\leq \int_{\xi}^{x} |f(t, y(t)) - f(t, z(t))| \, dt \\
&\leq \int_{\xi}^{x} L \cdot |y(t) - z(t)| \, dt \\
&= L \cdot \int_{\xi}^{x} |y(t) - z(t)| \cdot e^{-2L \cdot t} \cdot e^{2L \cdot t} dt \\
&\leq L \cdot \|y - z\|_w \cdot \int_{\xi}^{x} e^{2L \cdot t} dt \\
&\leq L \cdot \|y - z\|_w \cdot \tfrac{1}{2L} \cdot e^{2L \cdot x}.
\end{aligned}$$

It follows

$$|T(y)(x) - T(z)(x)| \cdot e^{-2L \cdot x} \leq \frac{L}{2L} \cdot \|y - z\|_w \quad \text{for all } x \in [\xi, \xi + a].$$

Hence,

$$\|T(y) - T(z)\|_w \leq \frac{1}{2} \cdot \|y - z\|_w.$$

Now, Banach's fixed point theorem can be applied, and therefore T has a unique fixed point, which is then the unique solution of the initial value problem.

The iteration $y^{(0)}(x) = \eta$ and

$$y^{(n+1)}(x) := T(y^{(n)})(x) = \eta + \int_{\xi}^{x} f(t, y^{(n)}(t))dt, \quad n = 0, 1, 2, 3, ...$$

is called Picard iteration. It converges to the unique solution. □

Remark 7.1. Banach's fixed point theorem is also called the Contraction-Mapping Theorem, especially when $E = \mathbb{R}^n$. It can also be used to prove the Inverse Function Theorem and the Implicit Function Theorem. See (68), for example. □

Remark 7.2. In (101) splitting methods were considered for solving linear systems of equations, say $Ax = b$, where $A \in \mathbb{R}^{n \times n}$ is nonsingular and $b \in \mathbb{R}^n$. Expressing the matrix A in the form $A = M - N$ with M being nonsingular is called a splitting of the matrix A.

Let x^* denote the unique solution of $Ax = b$, then we have

$$(M - N)x^* = b,$$

which can be written equivalently as

$$x^* = M^{-1}Nx^* + M^{-1}b.$$

Choosing $x^{(0)} \in \mathbb{R}^n$ arbitrarily we can start the fixed point iteration

$$x^{(n+1)} = M^{-1}N \cdot x^{(n)} + M^{-1}b, \quad n = 0, 1, 2, \dots . \tag{7.7}$$

Special choices of M and N lead to the well-known Jacobi, Gauss-Seidel, and the successive overrelaxation methods. Note that no fixed point theorem is needed, because the unique solution is guaranteed by the nonsingularity of A and we have

$$x^{(n+1)} - x^* = M^{-1}N \cdot (x^{(n)} - x^*) = \dots = (M^{-1}N)^n \cdot (x^{(0)} - x^*).$$

It can be shown (without any use of fixed point theory) that

$$\lim_{n\to\infty} (M^{-1}N)^n = O, \quad \text{(where } O \text{ denotes the zero matrix)},$$

if and only if the so-called spectral radius of $M^{-1}N$ is smaller than 1. So, the proof of convergence for these splitting methods reduces to the calculation of a spectral radius. Therefore, Theorem 7.2 is associated with these splitting methods only due the fixed point iteration (7.7), but not due to the proof of existence and uniqueness of a fixed point.

However, we refer to (7) where a generalization of these splitting methods to interval computations is done. There, the context to Theorem 7.2 is obvious and necessary. See also (8), (21), and Chapter 13 in (68). □

8 Schauder's Fixed Point Theorem

In this chapter we deal with the question if the Brouwer fixed point theorem can be extended from $\mathbb{R}^n$ to other normed linear spaces. First we consider n-dimensional normed linear spaces, second we consider infinite-dimensional normed linear spaces.

An n-dimensional normed linear space (different from $\mathbb{R}^n$) is presented in the following example.

Example 8.1. Let $f : [0,1] \to \mathbb{R}$ defined by $f(x) = 1$, let $g : [0,1] \to \mathbb{R}$ defined by $g(x) = x$, and let $h : [0,1] \to \mathbb{R}$ defined by $f(x) = x^2$. Then every quadratic function from $[0,1]$ to $\mathbb{R}$ can be expressed as a linear combination of f, g, and h. We may write

$$X = span\{f, g, h\}.$$

So, using the maximum norm $\|\cdot\|_\infty$ defined in Chapter 7.1 $(X, \|\cdot\|_\infty)$ is a three-dimensional normed linear space. For instance, any solution of the ordinary differential equation

$$y'''(x) = 0, \quad x \in [0,1]$$

belongs to X. □

As we will see in the following theorem any n-dimensional normed linear space and $\mathbb{R}^n$ are more or less the same.

Theorem 8.1. *Let $(X, \|\cdot\|_X)$ be an n-dimensional normed linear space. Then, X and $\mathbb{R}^n$ are isomorphic; i.e., there exists a linear bijection $T : X \to \mathbb{R}^n$ so that both T and T^{-1} are continuous.*

Proof: Let $f^{(1)}, f^{(2)}, ..., f^{(n)}$ be a basis of X. Then, for every $f \in X$ there exist uniquely defined $x_1, x_2, ..., x_n \in \mathbb{R}$ so that

$$f = \sum_{j=1}^{n} x_j \cdot f^{(j)}.$$

Defining $T : X \to \mathbb{R}^n$ by

$$T(f) = \begin{pmatrix} x_1 \\ \vdots \\ x_n \end{pmatrix}$$

we have a linear bijection with $T^{-1} : \mathbb{R}^n \to X$ defined by

$$T^{-1}\begin{pmatrix} x_1 \\ \vdots \\ x_n \end{pmatrix} = \sum_{j=1}^{n} x_j \cdot f^{(j)}.$$

To show continuity of T and T^{-1} we consider the norm $\|\cdot\|$ on $\mathbb{R}^n$ defined by

$$\|x\| := \|T^{-1}(x)\|_X \quad \text{where } x = \begin{pmatrix} x_1 \\ \vdots \\ x_n \end{pmatrix}.$$

By Theorem 2.2 this norm is equivalent to the Euclidean norm on $\mathbb{R}^n$. It follows that there exist positive numbers α and β so that

$$\alpha \cdot \|x\|_e \leq \|x\| \leq \beta \cdot \|x\|_e \quad \text{for all } x \in \mathbb{R}^n.$$

In other words, for $x, y \in \mathbb{R}^n$ and $f, g \in X$ we have

$$\|T^{-1}(x) - T^{-1}(y)\|_X = \|T^{-1}(x-y)\|_X = \|x-y\| \leq \beta \cdot \|x-y\|_e$$

and

$$\|T(f) - T(g)\|_e = \|T(f-g)\|_e \leq \frac{1}{\alpha} \cdot \|T(f-g)\| = \frac{1}{\alpha} \cdot \|f-g\|_X.$$

So, T as well as T^{-1} are (even Lipschitz) continuous. □

Concerning Brouwer's fixed point theorem we get the following consequence.

Corollary 8.1. *Let $(X, \|\cdot\|_X)$ be an n-dimensional normed linear space and let $K \subseteq X$ be nonempty, bounded, closed, and convex. Let $f : K \to X$ be continuous satisfying $f(K) \subseteq K$, then there exists $x^* \in K$ with $f(x^*) = x^*$.*

Proof: By Theorem 8.1 X and $\mathbb{R}^n$ are isomorphic; i.e., there exists a linear bijection $T : X \to \mathbb{R}^n$ so that both T and T^{-1} are continuous. Therefore, $\tilde{K} := T(K)$ is nonempty, bounded, closed (due to the continuity of T), and convex (due to the linearity of T) in $\mathbb{R}^n$. Defining

$$g := T \circ f \circ T^{-1}$$

we get $g : \tilde{K} \to \tilde{K}$ and we get that g is continuous. By Brouwer's fixed point theorem there exists $y^* \in \tilde{K}$ satisfying $g(y^*) = y^*$. Then, $x^* := T^{-1}(y^*) \in K$ is a fixed point of f. □

Brouwer's fixed point theorem is not true, if one just substitutes an infinite-dimensional normed linear space (even a very regular one) for $\mathbb{R}^n$. This will be shown in the following example.

Example 8.2. (Kakutani, 1943) Consider the space $X = l^2(\mathbb{Z})$ of all doubly-infinite real-valued sequences $\{x_n\}_{n=-\infty}^{\infty}$ ($\mathbb{Z}$ denotes the set of all integers) so that

$$\sum_{n=-\infty}^{\infty} x_n^2 < \infty;$$

equipped with the norm

$$\|x\|_2 := \sqrt{\sum_{n=-\infty}^{\infty} x_n^2}.$$

$(X, \|\cdot\|_2)$ is a Banach space. Furthermore, the closed unit ball

$$B = \{x \in X : \|x\|_2 \leq 1\}$$

is a nonempty, closed, bounded, and convex set in X. Let $S(x)$ denote the shifted sequence of x, that is

$$(S(x))_n = x_{n+1} \quad \text{for } n \in \mathbb{Z};$$

so, for instance, if $x = e_0$ is the sequence with 0's in each coordinate except the 0th coordinate where it has 1 for an entry, then $S(x) = e_1$, the sequence with a 1 in the 1th entry and 0's elsewhere.

It is plain and easy to see that S takes X onto itself in a linear, isometric manner:

$$\|S(x)\|_2 = \sqrt{\sum_{n=-\infty}^{\infty} x_{n+1}^2} = \sqrt{\sum_{n=-\infty}^{\infty} x_n^2} = \|x\|_2.$$

With Kakutani, define $\Phi : X \to X$ by

$$\Phi(x) = S(x) + \frac{1}{2}(1 - \|x\|_2) \cdot e_0.$$

Φ takes B into B: after all, if $\|x\|_2 \leq 1$, then

$$\begin{aligned} \|\Phi(x)\|_2 &\leq \|S(x)\|_2 + \|\tfrac{1}{2}(1 - \|x\|_2) \cdot e_0\|_2 \\ &= \|x\|_2 + \tfrac{1}{2}(1 - \|x\|_2) = \tfrac{1}{2} + \tfrac{1}{2} \cdot \|x\|_2 \leq 1. \end{aligned}$$

However, Φ is fixed point free in B. Indeed, imagine $x^* \in X$ satisfies $\Phi(x^*) = x^*$. Then

$$x^* - S(x^*) = \frac{1}{2}(1 - \|x^*\|_2) \cdot e_0. \tag{8.1}$$

Can $x^* = 0$, where 0 denotes the zero sequence? If so, then (8.1) says

$$0 = \frac{1}{2}(1 - \|0\|_2) \cdot e_0 = \frac{1}{2} \cdot e_0,$$

which is not so. So $x^* \neq 0$. But this says that

$$x^* - S(x^*) = \frac{1}{2}(1 - \|x^*\|_2) \cdot e_0$$

and by the very definition of $S(x^*)$ we have

$$x_n^* = x_{n+1}^* \quad \text{for any } n \neq 0.$$

This means that

$$... = x_{-2}^* = x_{-1}^* = x_0^*, \quad x_1^* = x_2^* = x_3^* = ...$$

Since $x^* \neq 0$, at least one of the numbers x_0^* and x_1^*, respectively, is not 0, in plain contradiction to $\sum\limits_{n=-\infty}^{\infty} (x_n^*)^2 < \infty$. □

To get a fixed point theorem in infinite-dimensional spaces one has to assume more. This was done by Juliusz Schauder and it will be presented in the following.

Before, we consider some properties of sets. Some of them depend on the dimension, some of them do not. For example, a set K is called to be convex, if it holds that

$$x, y \in K \quad \Rightarrow \quad \lambda \cdot x + (1 - \lambda) \cdot y \in K \quad \text{for all } \lambda \in (0, 1).$$

This definition does not depend on the dimension of K. Next, we consider open sets. Let $(E, \| \cdot \|)$ be a normed vector space and let $Q \subseteq E$. Q is said to be open, if for any $x \in Q$ there exists some $\varepsilon > 0$ so that

$$U(x; \varepsilon) := \{y \in E : \|x - y\| < \varepsilon\}$$

is a subset of Q. This definition is also independent from the dimension of E. So, let's start with some examples, where differences occur.

Example 8.3. Let $L = \{L_i\}_{i=1}^{\infty} \in l^2$ and let

$$\Omega = \{x \in l^2 : |x_i| < L_i, \, i = 1, 2, 3, ...\}.$$

On first sight one may think that Ω is an open set. But this is not the case as we will show now.

Let $\varepsilon > 0$ be arbitrary but fixed. Since $L \in l^2$, the sequence $\{L_i\}_{i=1}^{\infty}$ has to be convergent to 0; i.e., there exists $i_0 \in \mathbb{N}$ so that

$$L_i < \frac{\varepsilon}{4} \quad \text{for all } i \geq i_0 \, .$$

Let $x \in \Omega$. Then, we define $\xi = \{\xi_i\}_{i=1}^{\infty}$ by

$$\xi_i := \begin{cases} x_i, & i \neq i_0 \\ \dfrac{3 \cdot \varepsilon}{4} + x_{i_0}, & i = i_0. \end{cases}$$

Considering

$$\|x - \xi\| = \sqrt{\sum_{i=1}^{\infty} (x_i - \xi_i)^2} = \frac{3 \cdot \varepsilon}{4} < \varepsilon,$$

we can conclude that

$$\xi \in U(x; \varepsilon) := \{y \in l^2 : \|x - y\| < \varepsilon\}.$$

However, it holds

$$|\xi_{i_0}| = \left| \frac{3 \cdot \varepsilon}{4} + x_{i_0} \right| \geq \frac{3 \cdot \varepsilon}{4} - |x_{i_0}| \geq \frac{3 \cdot \varepsilon}{4} - \frac{\varepsilon}{4} = \frac{\varepsilon}{2} > L_{i_0},$$

which means that $\xi \notin \Omega$. Therefore, regardless of $\varepsilon > 0$

$$U(x; \varepsilon) \not\subset \Omega;$$

i.e., Ω is not an open set. □

A famous theorem by Bolzano-Weierstraß says that any sequence in a bounded subset of $\mathbb{R}^n$ has a convergent subsequence. This conclusion is not true in infinite-dimensional normed linear spaces as we will see in the following example.

Example 8.4. Let $(X, \|\cdot\|)$ be an infinite-dimensional normed linear space. We consider the unit ball

$$U(x_0; 1) := \{x \in X : \|x - x_0\| \leq 1\}.$$

We will show that there exists a sequence $\{x^{(n)}\}_{n=1}^{\infty}$ with $\|x^{(n)}\| = 1$, $n = 1, 2, 3, ...$ satisfying $\|x^{(m)} - x^{(k)}\| \geq 1$, if $m \neq k$. To achieve this, choose $x^{(1)} \in X$ so that $\|x^{(1)}\| = 1$ and define

$$X_1 := \{a \cdot x^{(1)} : a \in \mathbb{R}\}.$$

Since X is infinite-dimensional, there exists some $y \in X$, $y \notin X_1$ with

$$\|y - x\| > 0 \quad \text{for all } x \in X_1. \tag{8.2}$$

Defining the continuous function $\varphi : \mathbb{R} \to \mathbb{R}$ by

$$\varphi(a) = \|y - a \cdot x^{(1)}\|,$$

we have

$$\lim_{|a|\to\infty} \varphi(a) = \infty$$

since $\|y - a \cdot x^{(1)}\| \geq |a| \cdot \|x^{(1)}\| - \|y\|$. Therefore, there exists $a_1 \in \mathbb{R}$ with

$$\varphi(a_1) = \min_{a\in\mathbb{R}} \varphi(a) = \min_{a\in\mathbb{R}} \|y - a \cdot x^{(1)}\|. \tag{8.3}$$

In addition, $\|y - a_1 \cdot x^{(1)}\| > 0$ due to (8.2). Setting

$$x^{(2)} := \frac{1}{\|y - a_1 \cdot x^{(1)}\|} \cdot (y - a_1 \cdot x^{(1)})$$

we get $\|x^{(2)}\| = 1$ and

$$\begin{aligned}
\|x^{(2)} - x^{(1)}\| &= \left\| \frac{1}{\|y - a_1 \cdot x^{(1)}\|} \cdot (y - a_1 \cdot x^{(1)}) - x^{(1)} \right\| \\
&= \frac{1}{\|y - a_1 \cdot x^{(1)}\|} \cdot \left\| (y - a_1 \cdot x^{(1)}) - \|y - a_1 \cdot x^{(1)}\| \cdot x^{(1)} \right\| \\
&= \frac{1}{\|y - a_1 \cdot x^{(1)}\|} \cdot \left\| y - (a_1 + \|y - a_1 \cdot x^{(1)}\|) \cdot x^{(1)} \right\| \geq 1
\end{aligned}$$

thanks to (8.3). Now, suppose $x^{(1)}, ..., x^{(n)} \in X$ have been chosen so that

$$\|x^{(1)}\| = \|x^{(2)}\| = ... = \|x^{(n)}\| = 1$$

and

$$\|x^{(i)} - x^{(j)}\| \geq 1 \quad \text{for } i \neq j,\ i = 1, 2, ..., n,\ j = 1, 2, ..., n.$$

Then, we define

$$X_n := \{x = \sum_{j=1}^{n} t_j \cdot x^{(j)} : t_1, ..., t_n \in \mathbb{R}\}.$$

Since X is infinite-dimensional, there exists $z \in X$, but $z \notin X_n$ with

$$\|z - x\| > 0 \quad \text{for all } x \in X_n. \tag{8.4}$$

Again, we define a continuous function $\varphi : \mathbb{R}^n \to \mathbb{R}$ by

$$\varphi(t_1, ..., t_n) := \|z - \sum_{j=1}^{n} t_j \cdot x^{(j)}\|.$$

Then,

$$\|z-\sum_{j=1}^{n} t_j \cdot x^{(j)}\| = \|\sum_{j=1}^{n} t_j \cdot x^{(j)} - z\| = \|t_i \cdot x^{(i)} - (z - \sum_{j=1,\, j\neq i}^{n} t_j \cdot x^{(j)})\|.$$

Hence,

$$\varphi(t_1, ..., t_n) \geq |t_i| \cdot \|x^{(i)}\| - \|z - \sum_{j=1,\, j\neq i}^{n} t_j \cdot x^{(j)}\|$$

which implies that

$$\lim_{|t_i|\to\infty} \varphi(t_1, ..., t_n) = \infty \quad \text{for all } i \in \{1, ..., n\}. \tag{8.5}$$

Let $M > 0$ be arbitrary but fixed and let

$$K := \{t \in \mathbb{R}^n : \|t\|_e \leq M\}.$$

Then the function $\varphi : K \to \mathbb{R}$ has a minimum and a maximum, since K is compact and since φ is continuous. Together with (8.5) it follows that the function $\varphi : \mathbb{R}^n \to \mathbb{R}$ has a minimum, say

$$\varphi(t_1^*, ..., t_n^*) = \|z - \sum_{j=1}^{n} t_j^* \cdot x^{(j)}\| = \min_{(t_1,...,t_n)^T \in \mathbb{R}^n} \|z - \sum_{j=1}^{n} t_j \cdot x^{(j)}\| \tag{8.6}$$

and we have $\varphi(t_1^*, ..., t_n^*) > 0$ by (8.4). Setting

$$x^{(n+1)} := \frac{1}{\|z - \sum\limits_{j=1}^{n} t_j^* \cdot x^{(j)}\|} \cdot (z - \sum_{j=1}^{n} t_j^* \cdot x^{(j)})$$

we get $\|x^{(n+1)}\| = 1$ and for any $i \in \{1, ..., n\}$ we have

$$\|x^{(n+1)} - x^{(i)}\| = \left\| \frac{1}{\|z - \sum\limits_{j=1}^{n} t_j^* \cdot x^{(j)}\|} \cdot (z - \sum_{j=1}^{n} t_j^* \cdot x^{(j)}) - x^{(i)} \right\| =$$

$$\frac{1}{\|z - \sum\limits_{j=1}^{n} t_j^* \cdot x^{(j)}\|} \cdot \left\| z - \sum_{\substack{j=1,\\ j\neq i}}^{n} t_j^* \cdot x^{(j)} - \left(t_i^* + \|z - \sum_{j=1}^{n} t_j^* \cdot x^{(j)}\|\right) \cdot x^{(i)} \right\| \geq 1$$

thanks to (8.6). So, by induction, we have created a sequence $\{x^{(n)}\}_{n=1}^{\infty}$ with $\|x^{(n)}\| = 1$, $n = 1, 2, 3, ...$ satisfying

$$\|x^{(m)} - x^{(k)}\| \geq 1, \quad \text{if } m \neq k. \tag{8.7}$$

This sequence cannot have a convergent subsequence. □

8.1 Compactness

Definition 8.1. *Let X be a normed vector space. A nonempty set $K \subseteq X$ is called compact, if every sequence in K has a subsequence that converges to a limit that belongs to K.*

If $X = \mathbb{R}^n$, then the compact sets are precisely those that are closed and bounded; this is the celebrated theorem of Bolzano-Weierstraß. But, as we saw in Example 8.4, this is not so for the closed unit ball of an infinite-dimensional normed linear space, a closed, bounded set to be sure. Compactness is more special than being just closed and bounded; indeed, it is a key ingredient exploited by Schauder in his fixed point theorem.

We start by establishing a fundamental feature of compact sets in normed vector spaces - the property formulated is often referred to as 'total boundedness'.

Theorem 8.2. *Let X be a normed vector space and let $K \subseteq X$ be compact. Then, for every $\varepsilon > 0$ there exists a finite set of elements of K, say $\{v^{(1)}, ..., v^{(p)}\}$, so that for any $y \in K$ there exists $v^{(i)} \in \{v^{(1)}, ..., v^{(p)}\}$ with $\|v^{(i)} - y\| < \varepsilon$.*

Proof: Assume that K fails the conclusion. That is, assume, that there is some $\varepsilon_0 > 0$ so that for every finite set of elements of K, say $\{v^{(1)}, ..., v^{(n)}\}$, there exists some $y \in K$ with

$$\|v^{(j)} - y\| \geq \varepsilon_0 \quad \text{for all } v^{(j)} \in \{v^{(1)}, ..., v^{(n)}\}.$$

Let $x^{(1)} \in K$. Then, by assumption there is some $x^{(2)} \in K$ with

$$\|x^{(1)} - x^{(2)}\| \geq \varepsilon_0. \tag{8.8}$$

Considering the set $\{x^{(1)}, x^{(2)}\}$ by assumption there is some $x^{(3)} \in K$ with

$$\|x^{(j)} - x^{(3)}\| \geq \varepsilon_0 \quad \text{for } j = 1, 2.$$

Together with (8.8) we get

$$\|x^{(j)} - x^{(i)}\| \geq \varepsilon_0 \quad \text{for } j \neq i,\ j = 1, 2, 3,\ i = 1, 2, 3.$$

By induction, we get a sequence $\{x^{(n)}\}_{n=1}^{\infty} \subseteq K$ satisfying

$$\|x^{(j)} - x^{(i)}\| \geq \varepsilon_0 \quad \text{for } j \neq i,\ j = 1, 2, 3, ...,\ i = 1, 2, 3, ...$$

Such a sequence cannot have a convergent subsequence. As a result, K is not compact. □

Before we present Schauder's fixed point theorem in the next section, we want to give an important, classical example of a compact set in an infinite-dimensional normed linear space, the so-called Hilbert cube.

Example 8.5. Let $X = l^2$ and let $L = \{L_i\}_{i=1}^{\infty} \in l^2$. Then, the set

$$\Omega = \{x \in l^2 : |x_i| \leq L_i,\ i = 1, 2, 3, ...\}$$

is compact. To show this, let

$$\{x^{(n)}\}_{n=1}^{\infty} = \left\{\{x_i^{(n)}\}_{i=1}^{\infty}\right\}_{n=1}^{\infty}$$

be a sequence in Ω.

$i = 1 : \{x_1^{(n)}\}_{n=1}^{\infty} \subseteq [-L_1, L_1]$. Due to the theorem of Bolzano-Weierstraß there is a subsequence $\{x_1^{(\varphi_1(n))}\}_{n=1}^{\infty}$ and some $x_1^* \in [-L_1, L_1]$ with

$$\lim_{n\to\infty} x_1^{(\varphi_1(n))} = x_1^*.$$

$i = 2 :$

$$\left\{\begin{pmatrix} x_1^{(\varphi_1(n))} \\ x_2^{(\varphi_1(n))} \end{pmatrix}\right\}_{n=1}^{\infty} \subseteq \begin{pmatrix} [-L_1, L_1] \\ [-L_2, L_2] \end{pmatrix}.$$

Again, due to the theorem of Bolzano-Weierstraß there is a subsequence $\varphi_2(n)$ of $\varphi_1(n)$ and some $x_2^* \in [-L_2, L_2]$ with

$$\lim_{n\to\infty} x_2^{(\varphi_2(n))} = x_2^*.$$

So,

$$\lim_{n\to\infty} \begin{pmatrix} x_1^{(\varphi_2(n))} \\ x_2^{(\varphi_2(n))} \end{pmatrix} = \begin{pmatrix} x_1^* \\ x_2^* \end{pmatrix}.$$

Proceeding in this way we define some

$$x^* = \{x_i^*\}_{i=1}^{\infty} \in \Omega \tag{8.9}$$

and for any fixed k we have

$$\lim_{n\to\infty} \begin{pmatrix} x_1^{(\varphi_k(n))} \\ \vdots \\ x_k^{(\varphi_k(n))} \end{pmatrix} = \begin{pmatrix} x_1^* \\ \vdots \\ x_k^* \end{pmatrix}, \tag{8.10}$$

where

$$\varphi_k(n) \subseteq \varphi_{k-1}(n) \subseteq ... \subseteq \varphi_2(n) \subseteq \varphi_1(n) \subseteq \mathbb{N}. \tag{8.11}$$

Defining the diagonal subsequence $\{x^{(\varphi(n))}\}_{n=1}^{\infty}$ of $\{x^{(n)}\}_{n=1}^{\infty}$ by

$$\varphi(n) = \{\varphi_1(1),\, \varphi_2(2),\, \varphi_3(3),\, \varphi_4(4),\, ...\}$$

we get for any fixed k due to (8.11)

$$\begin{pmatrix} x_1^{(\varphi(n))} \\ \vdots \\ x_k^{(\varphi(n))} \end{pmatrix} \subseteq \begin{pmatrix} x_1^{(\varphi_k(n))} \\ \vdots \\ x_k^{(\varphi_k(n))} \end{pmatrix} \quad \text{for all } n \geq k. \tag{8.12}$$

Now, we will show that the sequence $\{x^{(\varphi(n))}\}_{n=1}^{\infty}$ converges to x^* from (8.9). Let $\varepsilon > 0$ be arbitrary, but fixed. Since $L = \{L_i\}_{i=1}^{\infty} \in l^2$, there exists some $k \in \mathbb{N}$ satisfying

$$\sqrt{\sum_{j=k+1}^{\infty} (2 \cdot L_j)^2} < \frac{\varepsilon}{2}.$$

Concerning this k, due to (8.10) and (8.12), there exists some $n_0 \in \mathbb{N}$ so that

$$\sqrt{\sum_{j=1}^{k} (x_j^{(\varphi(n))} - x_j^*)^2} < \frac{\varepsilon}{2} \quad \text{for all } n \geq \max\{n_0, k\}.$$

Then for all $n \geq \max\{n_0, k\}$ we get

$$\begin{aligned} \|x^{(\varphi(n))} - x^*\| &= \sqrt{\sum_{j=1}^{k} (x_j^{(\varphi(n))} - x_j^*)^2 + \sum_{j=k+1}^{\infty} (x_j^{(\varphi(n))} - x_j^*)^2} \\ &\leq \sqrt{\sum_{j=1}^{k} (x_j^{(\varphi(n))} - x_j^*)^2} + \sqrt{\sum_{j=k+1}^{\infty} (2 \cdot L_j)^2} < \frac{\varepsilon}{2} + \frac{\varepsilon}{2} = \varepsilon. \end{aligned}$$

That means, that $\{x^{(n)}\}_{n=1}^{\infty}$ has a convergent subsequence and its limit belongs to Ω by (8.9). Therefore, Ω is compact. □

Since continuity of a function $f : K \to K$ is not enough to ensure a fixed point in infinite-dimensional vector spaces, one has to modify the assumptions concerning f. The following definition will be sufficient.

Definition 8.2. *Let $(X, \|\cdot\|)$ be a normed vector space and $E \subseteq X$. A function (which is called an operator sometimes in infinite-dimensional spaces) $T : E \to X$ is called compact, if the following two conditions are fulfilled.*

1. *T is continuous (with respect to the given norm);*
2. *For any bounded $M \subseteq E$ the set $\overline{T(M)}$ is compact according to Definition 8.1, where $\overline{T(M)}$ denotes the closure of $T(M)$.*

Theorem 8.3. (Schauder's fixed point theorem) *Let X be a normed vector space and $K \subseteq X$ be a nonempty, convex, bounded, and closed set. Furthermore, let $T : K \to K$ be compact. Then, there exists some $x^* \in K$ satisfying $x^* = T(x^*)$.*

Proof: By Definition 8.2 the set $\overline{T(K)}$ is a compact set. Furthermore, since K is closed and since $T(K) \subseteq K$ we have

$$\overline{T(K)} \subseteq K.$$

Now, let $\varepsilon > 0$ be arbitrary but fixed. Then, because of Theorem 8.2 there exist $v^{(1)}, ..., v^{(p)} \in \overline{T(K)}$ so that for every $k \in K$ there is some $v^{(i)} \in \{v^{(1)}, ..., v^{(p)}\}$ satisfying $\|v^{(i)} - T(k)\| < \varepsilon$. Therefore, considering the functions

$$m_i(T(k)) := \left\{ \begin{array}{cl} \varepsilon - \|v^{(i)} - T(k)\| & \text{if } \|v^{(i)} - T(k)\| < \varepsilon, \\ 0 & \text{otherwise} \end{array} \right\} \quad i = 1, ..., p,$$

$m_i(T(k)) > 0$ for some i, $1 \leq i \leq p$, for each $k \in K$. So, the function $T_\varepsilon : K \to X$ defined by

$$T_\varepsilon(k) := \frac{1}{\sum\limits_{i=1}^{p} m_i(T(k))} \sum_{i=1}^{p} m_i(T(k)) \cdot v^{(i)}$$

is well defined. Since each $v^{(i)} \in \overline{T(K)} \subseteq K$ and since K is convex, it follows that $T_\varepsilon(K) \subseteq K$. Moreover, for every $k \in K$ it holds

$$\|T(k) - T_\varepsilon(k)\| \leq \frac{\sum\limits_{i=1}^{p} m_i(T(k)) \cdot \|T(k) - v^{(i)}\|}{\sum\limits_{i=1}^{p} m_i(T(k))} < \varepsilon. \tag{8.13}$$

Now, let $\{\varepsilon^{(n)}\}_{n=1}^{\infty}$ be a sequence in $\mathbb{R}$ satisfying

$$\varepsilon^{(n)} > 0 \quad \text{and} \quad \lim_{n\to\infty} \varepsilon^{(n)} = 0.$$

According to $\varepsilon^{(n)}$ we define $X_n := span\{v^{(1)}, ..., v^{(p_n)}\}$ and $K_n := K \cap X_n$. Then, by Theorem 8.1 it follows that X_n is isomorphic to $\mathbb{R}^n$ and obviously K_n is nonempty, convex, bounded, and closed. Furthermore, we have shown above that

$$T_{\varepsilon_n}(K_n) \subseteq K_n.$$

So, by Corollary 8.1 there exists $x^{(n)} \in K_n$ so that

$$T_{\varepsilon_n}(x^{(n)}) = x^{(n)}.$$

Since $T(x^{(n)}) \in T(K_n) \subseteq T(K)$ and since the set $\overline{T(K)}$ is compact, the sequence $\{T(x^{(n)})\}_{n=1}^{\infty}$ must have a convergent subsequence, say

$$\lim_{l \to \infty} T(x^{(n_l)}) = x^* \in \overline{T(K)} \subseteq K. \tag{8.14}$$

So, for every $\varepsilon > 0$ there is some $l_0 \in \mathbb{N}$ so that

$$\|x^* - T(x^{(n_l)})\| < \varepsilon \quad \text{for all } l > l_0.$$

According to (8.13), it holds

$$\|T(x^{(n_l)}) - T_{\varepsilon_{n_l}}(x^{(n_l)})\| < \varepsilon_{n_l} \quad \text{for all } l.$$

This results in

$$\begin{aligned} \|x^* - x^{(n_l)}\| &= \|x^* - T_{\varepsilon_{n_l}}(x^{(n_l)})\| \\ &\leq \|x^* - T(x^{(n_l)})\| + \|T(x^{(n_l)}) - T_{\varepsilon_{n_l}}(x^{(n_l)})\| \\ &< \varepsilon + \varepsilon_{n_l} \end{aligned}$$

for all $l > l_0$. In other words, we have

$$\lim_{l \to \infty} x^{(n_l)} = x^*.$$

Since T is continuous, we have

$$\lim_{l \to \infty} T(x^{(n_l)}) = T(x^*).$$

So, together with (8.14), we get $x^* = T(x^*)$. □

8.2 Peano's Existence Theorem

In 1890, G. Peano published the following result.

Theorem 8.4. *Let* $f : [\xi, \xi + a] \times [\eta - b, \eta + b] \to \mathbb{R}$ *be continuous. Then, by setting*

$$M := \max\{|f(x, y)| : (x, y) \in [\xi, \xi + a] \times [\eta - b, \eta + b]\} \quad \text{and} \quad \alpha := \min\{a, \frac{b}{M}\}$$

there exists a continuously differentiable function $y(x)$ *satisfying*

$$\begin{aligned} y'(x) &= f(x, y(x)), \quad x \in [\xi, \xi + \alpha] \\ y(\xi) &= \eta. \end{aligned}$$

The proof will be done via Schauder's fixed point theorem. Another auxiliary result will be the theorem of Ascoli-Arzela which is based on so-called equicontinuous sets of real-valued functions. So, we start with the definition of equicontinuity.

Definition 8.3. *Let* $F = \{f^{(1)}, f^{(2)}, ...\}$ *be a set of functions where* $f^{(i)} : [c, d] \to \mathbb{R}$ *are continuous for all* $f^{(i)} \in F$. *The set* F *is said to be equicontinuous, if for* $\varepsilon > 0$ *there exists some* $\delta = \delta(\varepsilon)$, *so that for all* $f^{(i)} \in F$ *it holds that*

$$|f^{(i)}(x) - f^{(i)}(\overline{x})| < \varepsilon \quad \text{if } |x - \overline{x}| < \delta.$$

Note, that for all $f^{(i)} \in F$ the same δ can be chosen. Obviously, if $F = \{f\}$; i.e., if F is a singleton, then F is equicontinuous if f is continuous. Consider the case that $F = \{f^{(1)}, f^{(2)}\}$ where both functions are continuous. So, for $\varepsilon > 0$ there exists some δ_1 so that

$$|f^{(1)}(x) - f^{(1)}(\overline{x})| < \varepsilon \quad \text{if } |x - \overline{x}| < \delta_1.$$

and there exists some δ_2 so that

$$|f^{(2)}(x) - f^{(2)}(\overline{x})| < \varepsilon \quad \text{if } |x - \overline{x}| < \delta_2.$$

Setting $\delta := \min\{\delta_1, \delta_2\}$ we have shown that F is equicontinuous. In the same way one can show that any finite set of continuous functions with the same domain is equicontinuous. Therefore, the question arises, if there are sets of continuous functions with the same domain that are not equicontinuous. This question will be answered in the following example.

Example 8.6. Let F be the set of all continuous functions with domain $D = [0,1]$ and let $\varepsilon = \frac{1}{2}$. Then, for every $\delta > 0$ we can define two different values $x_1 \in [0,1]$ and $x_2 \in [0,1]$ satisfying

$$0 < x_2 - x_1 < \delta$$

and based on these two values we can define a continuous function f satisfying $f(x_1) = 0$ and $f(x_2) = 1$. For example, f could be chosen as a piecewise linear function; i.e.,

$$f(x) = \begin{cases} 0 & \text{if } x \in [0, x_1], \\ \dfrac{x - x_1}{x_2 - x_1} & \text{if } x \in (x_1, x_2], \\ 1 & \text{if } x \in (x_2, 1]. \end{cases}$$

Then, we have

$$|x_2 - x_1| < \delta, \quad \text{but} \quad |f(x_2) - f(x_1)| = |1 - 0| = 1 \not< \frac{1}{2} = \varepsilon.$$

Hence, the set F is not equicontinuous. □

Equicontinuity plays an important role in the following theorem.

Theorem 8.5. (Theorem of Ascoli-Arzela) *Let $F = \{f^{(1)}, f^{(2)}, ...\}$ be equicontinuous, where $f^{(i)} : [c,d] \to \mathbb{R}$ for all $f^{(i)} \in F$. If there exists some $P \geq 0$ so that*

$$|f^{(i)}(x)| \leq P \quad \textit{for all } x \in [c,d], \quad \textit{and for all } f^{(i)} \in F,$$

then F has a uniformly convergent subsequence and its limit function is continuous.

Proof: Let $A = \{x_1, x_2, ...\}$ be the set of rational numbers in $[c,d]$. The sequence of real numbers $\{a_n\}_{n=1}^{\infty}$ with

$$a_n := f^{(n)}(x_1)$$

is bounded by P. The Bolzano-Weierstraß theorem tells us this sequence has a convergent subsequence, say

$$f^{(p_1)}(x_1),\ f^{(p_2)}(x_1),\ f^{(p_3)}(x_1) \ldots .$$

The sequence of real numbers $\{b_n\}_{n=1}^{\infty}$ with

$$b_n := f^{(p_n)}(x_2)$$

is bounded by P, too. Again, by the theorem of Bolzano-Weierstraß, this sequence has a convergent subsequence, say

$$f^{(q_1)}(x_2),\ f^{(q_2)}(x_2),\ f^{(q_3)}(x_2)\dots.$$

Here, $\{q_n\}_{n=1}^{\infty}$ is a subsequence of $\{p_n\}_{n=1}^{\infty}$. The third sequence $\{c_n\}_{n=1}^{\infty}$ with

$$c_n := f^{(q_n)}(x_3)$$

has a convergent subsequence, too, say

$$f^{(r_1)}(x_3),\ f^{(r_2)}(x_3),\ f^{(r_3)}(x_3)\dots.$$

Continuing this process we get several sequences

$$\begin{array}{lll} f^{(p_1)},\ f^{(p_2)},\ f^{(p_3)},\ f^{(p_4)},\ \dots & \text{convergent for} & x = x_1, \\ f^{(q_1)},\ f^{(q_2)},\ f^{(q_3)},\ f^{(q_4)},\ \dots & \text{convergent for} & x = x_1,\ x_2 \\ f^{(r_1)},\ f^{(r_2)},\ f^{(r_3)},\ f^{(r_4)},\ \dots & \text{convergent for} & x = x_1,\ x_2,\ x_3, \\ \dots\dots\dots\dots\dots\dots\dots & & \end{array}$$

The kth row is a subsequence of the $(k-1)$th row being convergent for $x = x_1, x_2, ..., x_k$. As a result, the diagonal sequence

$$\left.\begin{array}{c} \{f^{(d_n)}(x)\}_{n=1}^{\infty} := f^{(p_1)}(x),\ f^{(q_2)}(x),\ f^{(r_3)}(x),\ \dots \\ \text{is convergent for all } x \in A. \end{array}\right\} \tag{8.15}$$

Now, let $\varepsilon > 0$. Since $\{f^{(d_n)}(x)\}_{n=1}^{\infty}$ is equicontinuous, there exists $\delta(\varepsilon)$ so that for all $d_n \geq 1$ it holds that

$$|f^{(d_n)}(x) - f^{(d_n)}(\overline{x})| < \frac{\varepsilon}{3} \quad \text{if } |x - \overline{x}| < \delta(\varepsilon). \tag{8.16}$$

Then, we subdivide the interval $[c, d]$ in subintervals J_1, ...,J_p so that

$$[c, d] = J_1 \cup \dots \cup J_p \quad \text{with diameter of } J_i \leq \delta(\varepsilon), i = 1, ..., p.$$

For every J_i there exists x_i with

$$x_i \in J_i \cap A.$$

Furthermore, due to (8.15), there exists $n_0(\varepsilon)$ so that

$$|f^{(d_m)}(x_i) - f^{(d_n)}(x_i)| < \frac{\varepsilon}{3} \quad \text{for all } i = 1, ..., p \quad \text{if } m, n \geq n_o(\varepsilon). \tag{8.17}$$

As a result, if $x \in [c,d]$, there is some J_k with $x \in J_k$ and we get $|x - x_k| < \delta(\varepsilon)$, whence via (8.16) and (8.17) it follows

$$|f^{(d_m)}(x) - f^{(d_n)}(x)| \leq$$

$$|f^{(d_m)}(x) - f^{(d_m)}(x_k)| + |f^{(d_m)}(x_k) - f^{(d_n)}(x_k)| + |f^{(d_n)}(x_k) - f^{(d_n)}(x)| < \varepsilon$$

if $m, n \geq n_0(\varepsilon)$. Since $x \in [c,d]$ was arbitrary, then for $m, n \geq n_0(\varepsilon)$ we can conclude that

$$\varepsilon > \max\{|f^{(d_m)}(x) - f^{(d_n)}(x)| : x \in [c,d]\} = \|f^{(d_m)} - f^{(d_n)}\|_\infty.$$

This means, that $\{f^{(d_n)}\}_{n=1}^\infty$ is a Cauchy sequence subject to the maximum norm. But $(C([c,d]), \|\cdot\|_\infty)$ is a Banach space, so $\{f^{(d_n)}\}_{n=1}^\infty$ is convergent in this norm and the limit function is continuous. □

Now, we are able to prove Peano's existence theorem. Note that ξ, η, b, a, α, M, and the function f are given, there.

Proof of Theorem 8.4: We define the set of functions

$$K := \{w \in C([\xi, \xi + \alpha]) : w(x) = \eta + v(x),\ \|v\|_\infty \leq b\}.$$

Then, we define an operator T on K: for $w \in K$ we set

$$T(w)(x) := \eta + \int_\xi^x f(t, w(t))dt.$$

Since $f : [\xi, \xi + a] \times [\eta - b, \eta + b] \to \mathbb{R}$ is continuous, $T(w)(x)$ is continuous, too, and

$$\left| \int_\xi^x f(t, w(t))dt \right| \leq (x - \xi) \cdot M \leq \alpha \cdot M \leq \frac{b}{M} \cdot M = b$$

for all $x \in [\xi, \xi + \alpha] \subseteq [\xi, \xi + a]$. So, T is a self-mapping; i.e.,

$$T : K \to K.$$

Obviously, K is nonempty, convex, bounded, and closed. In order to apply Schauder's fixed point theorem, it remains to be shown that T is compact.

Firstly, we show, that T is continuous. Let

$$\lim_{n \to \infty} \|w^{(n)} - w^*\|_\infty = 0.$$

Then, since f is continuous,

$$\|T(w^{(n)}) - T(w^*)\|_\infty \le \int\limits_{\xi}^{\xi+\alpha} \|f(\cdot, w^{(n)}(\cdot)) - f(\cdot, w^*(\cdot))\|_\infty dt < \varepsilon$$

if n is sufficiently large. Secondly, let $\{w^{(n)}\}_{n=1}^{\infty}$ be a sequence in K. Since T is a self-mapping and since K is a bounded set, the set $\{T(w^{(n)})\}_{n=1}^{\infty}$ is bounded, and it only remains to show that $\{T(w^{(n)})\}_{n=1}^{\infty}$ is equicontinuous in order to apply the theorem of Ascoli-Arzela.

Let $\varepsilon > 0$ be given. Then, we choose

$$\delta := \frac{\varepsilon}{M}.$$

Hence, for every $n \ge 1$ and for every x, $\overline{x} \in [\xi, \xi + \alpha]$ with $|x - \overline{x}| < \delta$ we get

$$|T(w^{(n)})(x) - T(w^{(n)})(\overline{x})| \le \left| \int\limits_{\overline{x}}^{x} f(t, w^{(n)}(t))dt \right| \le |x - \overline{x}| \cdot M < \varepsilon.$$

Now, the theorem of Ascoli-Arzela can be applied, which means that T is compact. Then Schauder's fixed point theorem can be applied, which means that there exists some $\tilde{w} \in K$ satisfying $\tilde{w} = T(\tilde{w})$. It follows

$$\tilde{w}(x) = \eta + \int\limits_{\xi}^{x} f(t, \tilde{w}(t))dt$$

for $x \in [\xi, \xi + \alpha]$. Since this is equivalent to

$$\begin{aligned} \tilde{w}'(x) &= f(x, \tilde{w}(x)), \quad x \in [\xi, \xi + \alpha] \\ \tilde{w}(\xi) &= \eta \end{aligned}$$

Peano's existence theorem is proved. □

8.3 The Poincaré-Miranda Theorem in l^2

The following result is presented in this chapter, because its proof is similar to the proof of Schauder's fixed point theorem.

Theorem 8.6. *Let* $\hat{x} = \{\hat{x}_i\}_{i=1}^{\infty} \in l^2$, $L = \{l_i\}_{i=1}^{\infty} \in l^2$, $l_i \geq 0$ *for all* $i \in \mathbb{N}$, $\Omega := \{x \in l^2 : |x_i - \hat{x}_i| \leq l_i$ *for all* $i \in \mathbb{N}\}$ *and* $f : \Omega \to l^2$ *be a continuous function on* Ω. *Also let*

$$F_i^+ := \{x \in \Omega : x_i = \hat{x}_i + l_i\}, \quad F_i^- := \{x \in \Omega : x_i = \hat{x}_i - l_i\} \quad \textit{for all } i \in \mathbb{N}.$$

If for all $i \in \mathbb{N}$ *it holds that*

$$f_i(x) \cdot f_i(y) \leq 0 \quad \textit{for all } x \in F_i^+ \textit{ and for all } y \in F_i^-, \tag{8.18}$$

then there exists some $x^* \in \Omega$ *satisfying* $f(x^*) = 0$, *where* 0 *denotes the number zero as well as the zero sequence.*

Proof: For fixed $n \in \mathbb{N}$ we consider the function $\tilde{h}^{(n)} : \Omega \to l^2$ defined by

$$\tilde{h}^{(n)}(x) := \begin{pmatrix} f_1(x_1, x_2, ..., x_{n-1}, x_n, x_{n+1}, ...) \\ \vdots \\ f_n(x_1, x_2, ..., x_{n-1}, x_n, x_{n+1}, ...) \\ 0 \\ \vdots \end{pmatrix}.$$

As in Example 8.5 we can show that Ω is compact. Since f is continuous, the set $f(\Omega)$ is compact, too. Therefore, for given $\varepsilon > 0$ there is a finite set of elements $v^{(1)}, ..., v^{(p)} \in f(\Omega)$ so that if $f(x) \in f(\Omega)$, then there is a $v \in \{v^{(1)}, ..., v^{(p)}\}$ so that

$$\|f(x) - v\| \leq \varepsilon$$

and there exists $n_1 = n_1(\varepsilon) \in \mathbb{N}$ so that for all $n > n_1$ it holds that

$$\sqrt{\sum_{j=n+1}^{\infty} (v_j)^2} \leq \varepsilon \quad \text{for all } v \in \{v^{(1)}, ..., v^{(p)}\}.$$

So, if $n > n_1$ is valid, then for all $f(x) \in f(\Omega)$ we have some $v \in \{v^{(1)}, ..., v^{(p)}\}$ so that for all $x \in \Omega$

$$\|f(x) - \tilde{h}^{(n)}(x)\| = \| \begin{pmatrix} 0 \\ \vdots \\ 0 \\ f_{n+1}(x) \\ f_{n+2}(x) \\ \vdots \end{pmatrix} \| \leq \|f(x) - v\| + \| \begin{pmatrix} 0 \\ \vdots \\ 0 \\ v_{n+1} \\ v_{n+2} \\ \vdots \end{pmatrix} \| \leq 2\varepsilon.$$

Now, for fixed $n \in \mathbb{N}$ we define

$$\Omega_n := \begin{pmatrix} [\hat{x}_1 - l_1, \hat{x}_1 + l_1] \\ \vdots \\ [\hat{x}_n - l_n, \hat{x}_n + l_n] \end{pmatrix}$$

and $h^{(n)} : \Omega_n \to \mathbb{R}^n$ by

$$h^{(n)}(x) := \begin{pmatrix} f_1(x_1, x_2, ..., x_{n-1}, x_n, \hat{x}_{n+1}, \hat{x}_{n+2}, ...) \\ \vdots \\ f_n(x_1, x_2, ..., x_{n-1}, x_n, \hat{x}_{n+1}, \hat{x}_{n+2}, ...) \end{pmatrix}.$$

By (8.18) and Corollary 4.2 there exists $x^{(n)} \in \Omega_n$ with

$$h^{(n)}(x^{(n)}) = 0,$$

where 0 denotes the zero vector in $\mathbb{R}^n$. Setting

$$\tilde{x}^{(n)} := \begin{pmatrix} x^{(n)} \\ \hat{x}_{n+1} \\ \hat{x}_{n+2} \\ \vdots \end{pmatrix}$$

it holds that

$$\tilde{x}^{(n)} \in \Omega \quad \text{and} \quad \tilde{h}^{(n)}(\tilde{x}^{(n)}) = 0,$$

where 0 denotes the zero sequence. Now, let $n > n_1$. Then,

$$\|f(\tilde{x}^{(n)})\| = \|f(\tilde{x}^{(n)}) - \tilde{h}^{(n)}(\tilde{x}^{(n)})\| \leq 2\varepsilon.$$

Hence, $\lim_{n\to\infty} f(\tilde{x}^{(n)}) = 0$, where 0 denotes the zero sequence. Since Ω is compact, the sequence $\tilde{x}^{(n)}$ has a convergent subsequence with limit $x^* \in \Omega$. W.l.o.g. we assume that $\lim_{n\to\infty} \tilde{x}^{(n)} = x^*$ holds. On the one hand, it follows that $\lim_{n\to\infty} f(\tilde{x}^{(n)}) = f(x^*)$, since f is continuous. On the other hand, it follows that $f(x^*) = 0$, since the limit is unique. □

The infinite-dimensional Poincaré-Miranda theorem can be applied to difference equations that, in fact, are equivalent to some special infinite (denumerable) systems of equations, and to the controllability of nonlinear repetitive processes. See (47).

8.4 Verification Methods

In Chapter 6 verification methods were presented based on Brouwer's fixed point theorem. The problems, whose solutions have been verified there, were finite-dimensional. So, the idea is to apply Schauder's fixed point theorem to problems that are infinite-dimensional.

> But how can a computer, that can only represent finite many numbers, be used to verify solutions solving an infinite-dimensional problem?

We present the main ideas due to M. Plum by an illustrative example.

Example 8.7. Many boundary value problems for semilinear elliptic partial differential equations allow very stable numerical computations of approximate solutions, but are still lacking analytical existence proofs. Plum proposes a method which exploits the knowledge of a good numerical approximate solution, in order to provide a rigorous proof of existence of an exact solution close to the approximate one. This goal is achieved by Schauder's fixed point theorem as follows.

Consider the problem

$$\left.\begin{aligned} -(\triangle u)(x) + f(x, u(x)) &= 0, \quad x \in \Omega, \\ u(x) &= 0, \quad x \in \partial\Omega, \end{aligned}\right\} \tag{8.19}$$

where $\Omega \subset \mathbb{R}^n$, with $n \leq 3$, is a bounded domain with boundary $\partial\Omega$, $f : \overline{\Omega} \times \mathbb{R} \to \mathbb{R}$ is a given nonlinearity with f and $\frac{\partial f}{\partial y}$ being continuous, and $\triangle$ is the usual Laplacian; i.e., $\triangle = \sum\limits_{i=1}^{n} \frac{\partial^2}{\partial x_i^2}$.

Let $\tilde{u}(x)$ be an approximate solution of (8.19) calculated by some classical method (such as the finite element method). In order to verify, that at least in a neighbourhood of $\tilde{u}(x)$ there really exists a solution of (8.19), we consider the function set

$$\left.\begin{aligned} U = \tilde{u} + V, &\quad \text{where} \\ V = \{v \in C(\overline{\Omega}) : \|v\|_\infty \leq \alpha\} & \end{aligned}\right\} \tag{8.20}$$

with $\alpha > 0$ to be determined. Here, $C(\overline{\Omega})$ denotes the set of all real-valued continuous functions with domain $\overline{\Omega}$, and $\|v\|_\infty = \max\{|v(x)| : x \in \overline{\Omega}\}$.

> But how can a solution exist in U, although the Laplacian occurs in (8.19)?

The idea is to find a so-called 'weak solution'. To explain the concept of a weak solution, we need some definitions. We define

$$L^2(\Omega) = \{u : \Omega \to \mathbb{R} : u \text{ measurable }, \int\limits_\Omega u^2(x)dx < \infty\}$$

and $H := H^2(\Omega) \cap \mathring{H}^1(\Omega)$, consisting of all functions $u \in L^2(\Omega)$ which have weak derivatives $\frac{\partial u}{\partial x_i}$, $\frac{\partial^2 u}{\partial x_i \partial x_j}$ in $L^2(\Omega)$ (i.e., functions u for which functions

$$v_i =: \frac{\partial u}{\partial x_i} \in L^2(\Omega), \quad w_{ij} =: \frac{\partial^2 u}{\partial x_i \partial x_j} \in L^2(\Omega) \quad i, j = 1, ..., n$$

exist so that

$$\int\limits_\Omega u(x) \cdot \frac{\partial \varphi}{\partial x_i}(x)dx = -\int\limits_\Omega v_i(x) \cdot \varphi(x)dx \tag{8.21}$$

and

$$\int\limits_\Omega u(x) \cdot \frac{\partial^2 \varphi}{\partial x_i \partial x_j}(x)dx = \int\limits_\Omega w_{ij}(x) \cdot \varphi(x)dx \tag{8.22}$$

for all $\varphi \in C_0^\infty(\Omega))$[1], and which vanish on $\partial\Omega$ in the sense that

$$\|u - \varphi^{(n)})\|_{L^2(\Omega)} \to 0, \quad \|\frac{\partial u}{\partial x_i} - \frac{\partial \varphi^{(n)}}{\partial x_i}\|_{L^2(\Omega)} \to 0 \quad \text{as } n \to \infty \tag{8.23}$$

for some sequence $\{\varphi^{(n)}\}_{n=1}^\infty$ in $C_0^\infty(\Omega)$. Due to the embedding theorem of Sobolev-Kondrachev-Rellich (see (1)), we have indeed $H \subset C(\overline{\Omega}))$ and the embedding $H \hookrightarrow C(\overline{\Omega}))$ is compact (i.e., each $\|\cdot\|_H$-bounded sequence has a $\|\cdot\|_\infty$-convergent subsequence).

Now, we go back to our problem (8.19). Assume, the operator

$$L : H \to L^2(\Omega)$$

defined by

$$L(u)(x) := -(\triangle u)(x) + \frac{\partial f}{\partial y}(x, \tilde{u}(x)) \cdot u(x) \tag{8.24}$$

has an inverse

$$L^{-1} : L^2(\Omega) \to H$$

that is bounded. Then, we define the operator

$$T : C(\overline{\Omega}) \to C(\overline{\Omega})$$

[1] $C_0^\infty(\Omega)$ denotes the vector space of all infinitely differentiable functions on Ω with compact support in Ω.

by

$$T(u)(x) := L^{-1}\Big(\frac{\partial f}{\partial y}(x,\tilde{u}(x)) \cdot u(x) - f(x,u(x))\Big), \quad x \in \Omega.$$

Note, that $H \subset C(\overline{\Omega})$. If T had a fixed point, say u^*, then

$$u^*(x) = T(u^*)(x), \quad x \in \Omega$$

would lead to

$$L(u^*)(x) = \frac{\partial f}{\partial y}(x,\tilde{u}(x)) \cdot u^*(x) - f(x,u^*(x)), \quad x \in \Omega, \tag{8.25}$$

which, due to (8.24), is equivalent to

$$-(\triangle u^*)(x) + \frac{\partial f}{\partial y}(x,\tilde{u}(x)) \cdot u^*(x) = \frac{\partial f}{\partial y}(x,\tilde{u}(x)) \cdot u^*(x) - f(x,u^*(x)), \tag{8.26}$$

$x \in \Omega$; i.e.,

$$-(\triangle u^*)(x) + f(x,u^*(x)) = 0, \quad x \in \Omega. \tag{8.27}$$

Since $T(u^*) \in H$, we have that $u^* \in H$; i.e., u^* is a weak solution of (8.19). (The equalities of (8.25)-(8.27), and of $u^*(x) = 0$ on $\partial\Omega$ are only valid in the sense of (8.21), (8.22), and (8.23).)

To verify a fixed point of T, Schauder's fixed point theorem is applied based on the observation that T is compact, since the embedding $H \hookrightarrow C(\overline{\Omega})$ is compact. We have

$$\begin{aligned} T(u)(x) &= L^{-1}\Big(\tfrac{\partial f}{\partial y}(x,\tilde{u}(x)) \cdot u(x) - f(x,u(x))\Big) \\ &= \tilde{u}(x) - L^{-1}(L(\tilde{u}))(x) + L^{-1}\Big(\tfrac{\partial f}{\partial y}(x,\tilde{u}(x)) \cdot u(x) - f(x,u(x))\Big) \\ &\overset{(8.24)}{=} \tilde{u}(x) - L^{-1}\Big(-(\triangle\tilde{u})(x) + \frac{\partial f}{\partial y}(x,\tilde{u}(x)) \cdot (\tilde{u}(x) - u(x)) + f(x,u(x))\Big). \end{aligned}$$

Setting

$$g(x,y) := f(x,y) - f(x,\tilde{u}(x)) - \frac{\partial f}{\partial y}(x,\tilde{u}(x)) \cdot (y - \tilde{u}(x))$$

we get by Taylor's formula

$$\frac{g(x,y)}{|y - \tilde{u}(x)|} \to 0 \quad \text{as } y \to \tilde{u}(x) \tag{8.28}$$

and

$$f(x,\tilde{u}(x)) + g(x,u(x)) = f(x,u(x)) + \frac{\partial f}{\partial y}(x,\tilde{u}(x)) \cdot (\tilde{u}(x) - u(x)).$$

So, in order to apply Schauder's fixed point theorem we have to find α that satisfies

$$\| - L^{-1}\Big(- \triangle\tilde{u} + f(\cdot, \tilde{u}) + g(\cdot, \tilde{u} + v)\Big)\|_\infty \leq \alpha \tag{8.29}$$

for all $v \in C(\overline{\Omega})$ with $\|v\|_\infty \leq \alpha$. Then, we have (see (8.20))

$$T(U) \subseteq U, \tag{8.30}$$

and we can conclude that in U there is a weak solution of (8.19).

With (8.28) in hand it is not unlikely that (8.29) can be verified, if $\tilde{u}$ is a good approximation (and if there really exists a solution).

In order to verify (8.29), operator norm bounds for L^{-1}, $-\triangle\tilde{u} + f(\cdot, \tilde{u})$, and g are needed. Plum calculated eigenvalue bounds for L of (8.24) and bounds for an integral with known integrand. Interval arithmetic was used in all numerical evaluations (but not during the computation of the approximate solution $\tilde{u}$!), in order to take rounding errors into account on a computer.

We do not go into further details, but we present a numerical example published by Breuer, McKenna, and Plum in 2003. Let $\Omega = (0,1) \times (0,1)$ and $x = \binom{x_1}{x_2}$. They considered the problem

$$\left.\begin{aligned} -(\triangle u)(x) - (u(x))^2 &= -s \cdot \sin(\pi \cdot x_1) \cdot \sin(\pi \cdot x_2), \quad x \in \Omega, \\ u(x) &= 0, \quad x \in \partial\Omega. \end{aligned}\right\} \tag{8.31}$$

It had been conjectured in the PDE community since the 1980's that problem (8.31) has at least 4 solutions for $s > 0$ sufficiently large. For $s = 800$, Breuer, McKenna, and Plum were able to compute 4 essentially different approximate solutions by the numerical mountain pass algorithm developed in (20), where "essentially different" means that none of them is an elementary symmetry transform of another one. With Plum's method they could verify, by use of interval arithmetic exploited on a computer, that in a neighborhood of each of these 4 approximate solutions, indeed there really exists a solution.

Two years after publication, Dancer and Yan gave a more general analytical proof. They even proved that the number of solutions of problem (8.31) becomes unbounded as $s \to \infty$. Nevertheless, Breuer, McKenna, and Plum were first. □

Other people have also considered so-called enclosure methods for the problem (8.19) using Schauder's fixed point theorem. We want to mention M. T. Nakao and his co-workers. Their method avoids the computation of a bound for L^{-1}. On the other hand, it requires the verified solution of large nonlinear and linear systems, where the latter moreover have an interval right-hand side, and it requires explicit knowledge of projection error bounds. We refer to (65).

9 Tarski's Fixed Point Theorem

Continuity was the starting point to guarantee the existence of a fixed point:

$$\left.\begin{array}{l} f:[a,b]\to[a,b], \\ f \text{ is continuous} \end{array}\right\} \Rightarrow \left\{\begin{array}{c} \text{there exists } x^*\in[a,b] \\ \text{with } x^*=f(x^*). \end{array}\right.$$

We have presented several extensions. For example, $f : K \to K$, where $K \subseteq \mathbb{R}^n$ is convex, bounded, and closed. Then, we have weakened the assumption that f has to be a self-mapping, and Schauder's fixed point theorem considers compact mappings in infinite-dimensional vector spaces.

All these results exploit the fact that f is continuous. So, on first sight, it seems implausible to get a fixed point theorem that does not assume continuity. However, let us consider the following situation:

$$f:[a,b]\to[a,b],\ f \text{ is increasing;}$$

i.e., if $x \leq y$, then $f(x) \leq f(y)$. Figure 9.1 shows that such a function must have a fixed point, although the function is not continuous. It was A. Tarski, who followed this idea. The starting point is no longer a vector space, but it is an ordered set (in order to generalize what is meant by increasing).

Definition 9.1. *Let Ω be a nonempty set. Then, $(\Omega, \leq)$ is called an ordered set, if for all x, y, $z \in \Omega$*

$$\begin{array}{rcl} x & \leq & x, \\ x\leq y,\, y\leq x & \Rightarrow & x=y \\ x\leq y,\, y\leq z & \Rightarrow & x\leq z. \end{array}$$

Example 9.1. We give some examples of ordered sets.
a) $(\mathbb{R}^n, \leq)$ with

$$\begin{pmatrix} x_1 \\ \vdots \\ x_n \end{pmatrix} \leq \begin{pmatrix} y_1 \\ \vdots \\ y_n \end{pmatrix} \quad :\Leftrightarrow \quad x_i \leq y_i \quad \text{for } i=1,\dots,n.$$

b) $(l^2, \leq)$, $(c_0, \leq)$, and $(l^\infty, \leq)$ with

$$\{x_n\}_{n=1}^\infty \leq \{y_n\}_{n=1}^\infty \quad :\Leftrightarrow \quad x_n \leq y_n \quad \text{for } n = 1, 2, 3, \ldots .$$

c) $(C([a,b]), \leq)$ with

$$f \leq g \quad :\Leftrightarrow \quad f(x) \leq g(x) \quad \text{for all } x \in [a,b].$$

To formulate a fixed point theorem, we need only one more definition.

Definition 9.2. *An ordered set $(\Omega, \leq)$ is said to be a complete lattice, if for any nonempty $\Lambda \subseteq \Omega$*

$$\inf \Lambda \text{ as well as } \sup \Lambda$$

exist and

$$\inf \Lambda \in \Omega \text{ as well as } \sup \Lambda \in \Omega.$$

Now, we can present Tarski's fixed point theorem.

Theorem 9.1. (Tarski's fixed point theorem, 1955) *Let $(\Omega, \leq)$ be a complete lattice and let $\Phi : \Omega \to \Omega$ be a function satisfying*

$$\omega \leq v \quad \Rightarrow \quad \Phi(\omega) \leq \Phi(v), \quad \text{for all } \omega, v \in \Omega.$$

Then, there exists some $\omega^* \in \Omega$ satisfying $\omega^* = \Phi(\omega^*)$.

Proof: Since $(\Omega, \leq)$ is a complete lattice, we have that $\inf \Omega \in \Omega$ as well as $\sup \Omega \in \Omega$ exist. So, $\sup \Omega = \max \Omega$ and $\inf \Omega = \min \Omega$.

Moreover, since Φ is an increasing self-mapping, we have

$$\min \Omega \leq \Phi(\min \Omega) \leq \Phi(\max \Omega) \leq \max \Omega.$$

Motivated by Figure 9.1 we define

$$\Lambda := \{\omega \in \Omega : \omega \leq \Phi(\omega)\}.$$

We know that $\Lambda \neq \emptyset$, because $\min \Omega \in \Lambda$. Again, since $(\Omega, \leq)$ is a complete lattice, the existence of $\omega_0 := \sup \Lambda$ is guaranteed. This means, that

$$\omega \leq \omega_0 \quad \text{for all } \omega \in \Lambda.$$

Since Φ is increasing by the definition of Λ we get

$$\omega \leq \Phi(\omega) \leq \Phi(\omega_0) \quad \text{for all } \omega \in \Lambda.$$

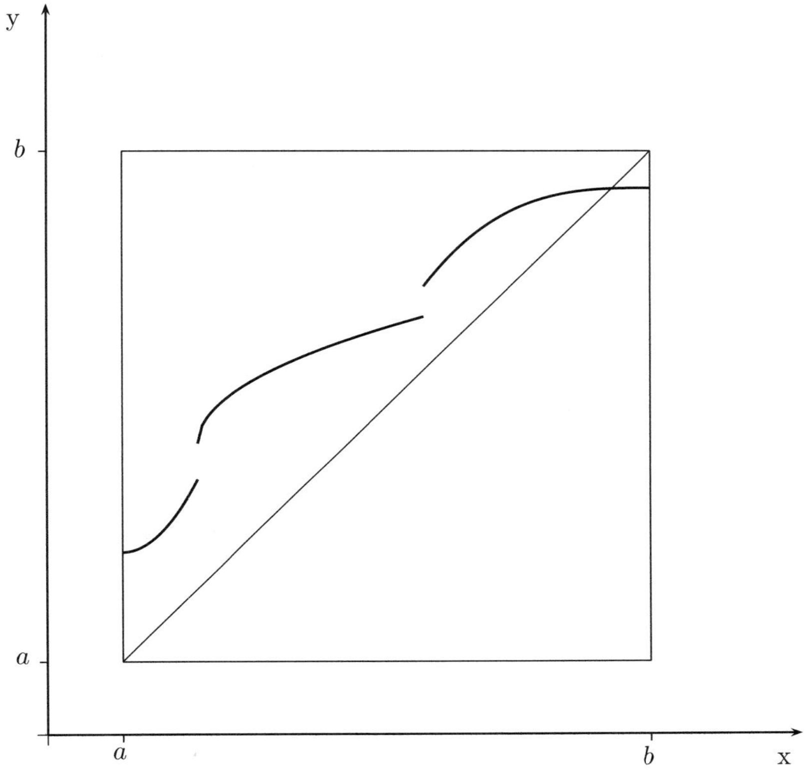

Figure 9.1: **There must be a fixed point, only due to the fact, that f is an increasing self-mapping**

So, $\Phi(\omega_0)$ is an upper bound of Λ. According to the definition of the supremum of a set we get

$$\omega_0 \leq \Phi(\omega_0), \tag{9.1}$$

whence it follows that $\Phi(\omega_0) \leq \Phi(\Phi(\omega_0))$ by the monotonicity of Φ. One can conclude that $\Phi(\omega_0) \in \Lambda$ and by the very definition of the supremum of a set we get

$$\Phi(\omega_0) \leq \omega_0.$$

Together with (9.1) we finally get $\Phi(\omega_0) = \omega_0$. □

As the reader might have noticed, the proof is very short and elementary. How can it be applied? First, we need to get a hint at what we're dealing with - what are some pertinent examples of complete lattices?

Remembering Figure 5.1 at page 48, we know that $\mathbb{R}^n$ with $n \geq 2$ is not a complete lattice. However, the next example will present a complete lattice. On first sight, this example looks strange, but it will be used in the following sections.

Example 9.2. Let $T > 0$ be given. Furthermore, let $\Gamma \geq 0$, $\tilde{L} \geq 0$, and $L^* \geq 0$ be given. Then, we consider $\Omega = \Omega(\Gamma, \tilde{L}, L^*)$ with

$$\begin{aligned} \Omega \quad := \quad & \Big\{ \omega = \{\omega_n\}_{n=1}^{\infty} : \mathbb{R} \times [0,T] \to l^{\infty}, \\ & \text{for all } n \in \mathbb{N} \text{ it holds } |\omega_n(x,t)| \leq \Gamma \text{ for all } (x,t) \in \mathbb{R} \times [0,T], \\ & \text{and for all } (x,t),(y,s) \in \mathbb{R} \times [0,T] \text{ it holds} \\ & \|\omega(x,t) - \omega(y,s)\| \leq \tilde{L} \cdot |x-y| + L^* \cdot |t-s| \Big\}. \end{aligned}$$

I.e., Ω is a set of l^{∞}-valued, bounded, Lipschitz continuous functions, and we want to show that $(\Omega, \leq)$ is a complete lattice, where for ω, $v \in \Omega$ it is

$$\omega \leq v \quad :\Leftrightarrow \quad \begin{cases} \text{for all } n = 1,2,3,\ldots \text{ it holds that } \omega_n(x,t) \leq v_n(x,t) \\ \text{for all } (x,t) \in \mathbb{R} \times [0,T]. \end{cases}$$

Of course, Ω is nonempty, since the zero element of l^{∞}, interpreted as the constant sequence of the zero function, belongs to Ω. Now, let $\Lambda \subseteq \Omega$ and let Λ be nonempty. Then, we have to show, that $\inf \Lambda$ as well as $\sup \Lambda$ exist and that $\inf \Lambda \in \Omega$ as well as $\sup \Lambda \in \Omega$ is fulfilled. By the definition of Ω, we have for all $\omega \in \Omega$ (and therefore, for all $\omega \in \Lambda$)

$$|\omega_n(x,t)| \leq \Gamma \quad \text{for all } n = 1,2,3,\ldots \text{ and for all } (x,t) \in \mathbb{R} \times [0,T]. \tag{9.2}$$

So, for any $n \in \mathbb{N}$ and for any $(x,t) \in \mathbb{R} \times [0,T]$ we define

$$\tilde{\omega}_n(x,t) := \sup\{\omega_n(x,t) : \omega \in \Lambda\}. \tag{9.3}$$

By definition (see (9.2)) $|\tilde{\omega}_n(x,t)| \leq \Gamma$ for all $n \in \mathbb{N}$ and for all $(x,t) \in \mathbb{R} \times [0,T]$. It remains to show that

$$\|\tilde{\omega}(x,t) - \tilde{\omega}(y,s)\| \leq \tilde{L} \cdot |x-y| + L^* \cdot |t-s| \text{ for all } (x,t),(y,s) \in \mathbb{R} \times [0,T].$$

So, let $(x,t),(y,s) \in \mathbb{R} \times [0,T]$, and let $n \in \mathbb{N}$ be arbitrary but fixed. Then for all $\omega \in \Lambda$

$$\omega_n(x,t) - \omega_n(y,s) \leq |\omega_n(x,t) - \omega_n(y,s)| \leq \tilde{L} \cdot |x-y| + L^* \cdot |t-s|$$

and it follows that

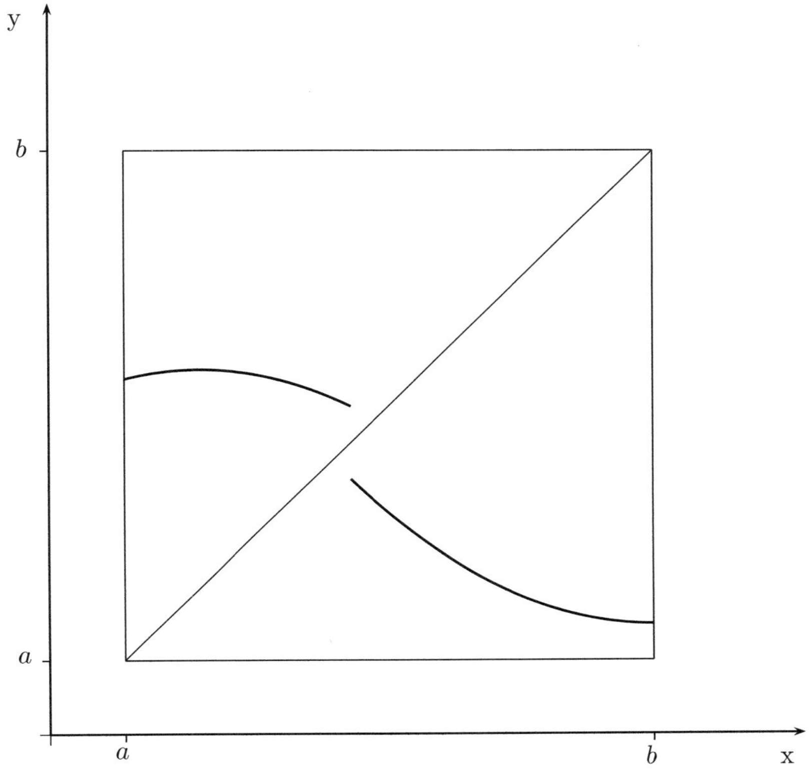

Figure 9.2: **No fixed point**

$$\omega_n(x,t) \le \tilde{L} \cdot |x-y| + L^* \cdot |t-s| + \omega_n(y,s) \le \tilde{L} \cdot |x-y| + L^* \cdot |t-s| + \tilde{\omega}_n(y,s).$$

In other words,

$$\tilde{L} \cdot |x-y| + L^* \cdot |t-s| + \tilde{\omega}_n(y,s)$$

is an upper bound for $\omega_n(x,t)$, $\omega \in \Lambda$. By definition of the supremum we get

$$\tilde{\omega}_n(x,t) \le \tilde{L} \cdot |x-y| + L^* \cdot |t-s| + \tilde{\omega}_n(y,s).$$

As a result

$$\tilde{\omega}_n(x,t) - \tilde{\omega}_n(y,s) \le \tilde{L} \cdot |x-y| + L^* \cdot |t-s|.$$

Exchanging (x,t) with (y,s) one can see that

$$\tilde{\omega}_n(y,s) - \tilde{\omega}_n(x,t) \le \tilde{L} \cdot |y-x| + L^* \cdot |s-t|.$$

That is

$$|\tilde{\omega}_n(x,t) - \tilde{\omega}_n(y,s)| \le \tilde{L} \cdot |x-y| + L^* \cdot |t-s|.$$

Since the right-hand side does not depend on n, we have

$$\|\tilde{\omega}(x,t) - \tilde{\omega}(y,s)\| \leq \tilde{L} \cdot |x-y| + L^* \cdot |t-s|.$$

It follows that $\tilde{\omega} \in \Omega$, and it is clear from (9.3) that $\tilde{\omega} = \sup \Lambda$. In the same way, one can show that $\inf \Lambda$ exists and that $\inf \Lambda \in \Omega$. □

Finally, we want to note that Tarski's fixed point theorem is no longer true, if one substitutes

$$\omega \leq v \quad \Rightarrow \quad \Phi(\omega) \leq \Phi(v)$$

by

$$\omega \leq v \quad \Rightarrow \quad \Phi(\omega) \geq \Phi(v).$$

See Figure 9.2.

9.1 Differential Equations in Banach Spaces

In 1950, Dieudonné published an example that shows that Peano's existence theorem is no longer true, if the considered function f is not real-valued, but has its values in a real Banach space. More precisely, let E be a real Banach space, let $a \in E$, and let $T > 0$. Then, we consider the so-called Cauchy problem

$$\left.\begin{array}{rcl} u'(t) & = & f(t, u(t)), \quad t \in [0,T], \\ u(0) & = & a, \end{array}\right\} \tag{9.4}$$

where

$$f : [0,T] \times E \to E$$

is continuous and bounded. The following example shows, that, in general, this problem has no solution.

Example 9.3. (Dieudonné, 1950) Let $E = c_0$, let $0 < T \leq 2$, and let

$$a = \left\{1, \frac{1}{2}, \frac{1}{3}, \frac{1}{4}, \ldots\right\} \in c_0.$$

Furthermore, let $f : [0,T] \times c_0 \to c_0$ be defined by

$$f\Big(t, \{x_n\}_{n=1}^{\infty}\Big) = \Big\{f_m\Big(t, \{x_n\}_{n=1}^{\infty}\Big)\Big\}_{m=1}^{\infty}$$

with

$$f_m\Big(t, \{x_n\}_{n=1}^{\infty}\Big) = \left\{\begin{array}{cl} 0 & \text{if } x_m \leq 0, \\ \sqrt{x_m} & \text{if } 0 \leq x_m \leq 4, \\ 2 & \text{if } 4 \leq x_m. \end{array}\right.$$

I.e., the functions f_m are independent of t and we have

$$\lim_{m\to\infty} f_m\Big(t, \{x_n\}_{n=1}^{\infty}\Big) = 0 \quad \text{if} \quad \lim_{m\to\infty} x_m = 0.$$

Therefore, we have $f : [0,T] \times c_0 \to c_0$. If (9.4) had a solution $u : [0,T] \to c_0$; i.e.,

$$u(t) = \{u_1(t), u_2(t), u_3(t), ...\},$$

it would follow

$$u'_m(t) = \left\{ \begin{array}{cl} 0 & \text{if } u_m(t) \le 0, \\ \sqrt{u_m(t)} & \text{if } 0 \le u_m(t) \le 4, \\ 2 & \text{if } 4 \le u_m(t), \end{array} \right\} \quad t \in [0,T], \quad u_m(0) = \frac{1}{m}$$

for all $m \in \mathbb{N}$. For fixed $m \in \mathbb{N}$ we get the (unique) solution

$$u_m(t) = \left(\frac{t}{2} + \frac{1}{\sqrt{m}}\right)^2, \quad t \in [0,T].$$

(Note, that $T \le 2$.) However, due to

$$\lim_{m\to\infty} u_m(t) = \frac{t^2}{4} > 0 \quad \text{for all } t \in (0,T],$$

we would have

$$u(t) = \{u_1(t), u_2(t), u_3(t), ...\} \notin c_0.$$

Therefore, the problem (9.4) has no solution in this situation. □

Godunov (1974) showed that whenever E is an infinite-dimensional Banach space there exist an a in E, a $T > 0$, and a continuous f so that the problem (9.4) has no solution. So, a new field in mathematics was born: Differential Equations in Banach Spaces, dealing with the question what conditions have to be fulfilled so that a solution of (9.4) can be guaranteed. We refer to the book of K. Deimling and the paper of P. Volkmann for a detailed introduction.

In the following section, we will present an existence theorem for a parabolic differential equation in l^∞ based on Tarski's fixed point theorem. Its advantage is that there is no need to consider compact functions as it is the case in Schauder's fixed point theorem, so we are able to get an existence theorem in l^∞.

9.2 Applying Tarski's Fixed Point Theorem

We consider the parabolic differential equation

$$\left.\begin{aligned} u_t(x,t) - u_{xx}(x,t) &= f(x,t,u(x,t)), & (x,t) \in \mathbb{R} \times (0,T), \\ u(x,0) &= 0, & x \in \mathbb{R}, \end{aligned}\right\} \tag{9.5}$$

with

$$f : \mathbb{R} \times (0,T) \times l^\infty \to l^\infty.$$

I.e., the values of a solution $u(x,t)$ have to be in l^∞. Before we attack this problem, we need to explain what is meant by $u_t(x,t)$ and $u_{xx}(x,t)$.

Definition 9.3. *Let E be a Banach space. Then, a function $u : \mathbb{R} \times (0,T) \to E$ is partially differentiable with respect to t in $(x_0,t_0) \in \mathbb{R} \times (0,T)$, if there exists*

$$u_t(x_0,t_0) \in E$$

so that

$$\lim_{t \to t_0} \left\| \frac{1}{t-t_0} \cdot \Big(u(x_0,t) - u(x_0,t_0) \Big) - u_t(x_0,t_0) \right\| = 0$$

holds.

Analogously, one can define $u_x(x,t)$ and $u_{xx}(x,t)$.

If the function $f(x,t,z)$ in (9.5) does not depend on z, the problem (9.5) can be solved by the following theorem.

Theorem 9.2. *Let E be any real Banach space and let $g(x,t) : \mathbb{R} \times [0,T] \to E$ be bounded, continuous, and Lipschitz continuous in $x \in \mathbb{R}$, uniformly with respect to t; i.e., there exists a constant L so that*

$$\|g(x,t) - g(y,t)\| \le L \cdot |x-y|$$

for all $(x,t), (y,t) \in \mathbb{R} \times [0,T]$. Then, the function

$$u(x,t) := \int_0^t \int_{-\infty}^{\infty} \frac{1}{\sqrt{4\pi(t-\tau)}} \cdot e^{-\dfrac{(x-\xi)^2}{4(t-\tau)}} \cdot g(\xi,\tau)\, d\xi\, d\tau$$

is a solution of the Cauchy problem

$$\begin{aligned} u_t(x,t) - u_{xx}(x,t) &= g(x,t), & (x,t) \in \mathbb{R} \times (0,T) \\ u(x,0) &= 0, & x \in \mathbb{R}. \end{aligned}$$

In addition, $u(x,t)$ is continuous on $\mathbb{R} \times [0,T]$.

For a proof we refer to (33), pp. 1-25. There, the theorem is only given for real-valued functions, but the theorem remains true if $g(x,t)$ has got its values in any real Banach space E, because one has just to substitute the norm for the absolute value. Also, if $h : [a,b] \to E$ is continuous, then

$$\int_a^b h(x)dx \quad \text{as well as} \quad \int_{-\infty}^{\infty} h(x)dx$$

have to be understood as limits of Riemann sums $\sum h(\xi_k) \cdot (x_k - x_{k-1})$. Now, we can attack problem (9.5).

Theorem 9.3. *Let $f : \mathbb{R} \times [0,T] \times l^\infty \to l^\infty$ be a function with the following properties:*

i) f is continuous.

ii) There exists a constant L_1 so that for all $(x,t,z), (y,t,z) \in \mathbb{R} \times [0,T] \times l^\infty$ it holds

$$\|f(x,t,z) - f(y,t,z)\| \le L_1 \cdot |x - y|.$$

iii) There exists a constant L_2 so that for all $(x,t,z), (x,t,\tilde{z}) \in \mathbb{R} \times [0,T] \times l^\infty$ it holds

$$\|f(x,t,z) - f(x,t,\tilde{z})\| \le L_2 \cdot \|z - \tilde{z}\|.$$

iv) There exists a constant M so that for all $(x,t,z) \in \mathbb{R} \times [0,T] \times l^\infty$ it holds

$$\|f(x,t,z)\| \le M.$$

v) For all (x,t,z), $(x,t,\tilde{z}) \in \mathbb{R} \times [0,T] \times l^\infty$ it holds

$$z \le \tilde{z} \quad \Rightarrow \quad f(x,t,z) \le f(x,t,\tilde{z}).$$

Then, there exists a continuous function $u : \mathbb{R} \times [0,T] \to l^\infty$ satisfying (9.5).

Proof: We consider the following set of functions

$$\Omega := \Big\{\omega : \mathbb{R} \times [0,T] \to l^\infty, \ -\Gamma \le \omega \le \Gamma$$
$$\|\omega(x,t) - \omega(y,s)\| \le L_3 \cdot |x - y| + L_4 \cdot |t - s|,$$
$$\text{for all } (x,t),(y,s) \in \mathbb{R} \times [0,T]\Big\},$$

with

$$\Gamma = \{\Gamma_n\}_{n=1}^{\infty} \quad \text{and} \quad \Gamma_n = T \cdot M,$$

$L_3 := 2 \cdot M \cdot \sqrt{T}$ and $L_4 := M + 2(L_1 + L_2 \cdot L_3) \cdot \sqrt{\frac{T}{\pi}}$. Moreover, for $\omega \in \Omega$ we define

$$\Phi(\omega)(x,t) := \int_0^t \int_{-\infty}^{\infty} \frac{1}{\sqrt{4\pi(t-\tau)}} \cdot e^{-\dfrac{(x-\xi)^2}{4(t-\tau)}} \cdot f(\xi,\tau,\omega(\xi,\tau))\, d\xi\, d\tau.$$

Then, for $g(x,t) := f(x,t,\omega(x,t))$ we have

$$\begin{aligned} \|g(x,t) - g(y,t)\| &= \|f(x,t,\omega(x,t)) - f(y,t,\omega(y,t))\| \\ &\leq \|f(x,t,\omega(x,t)) - f(y,t,\omega(x,t))\| + \\ &\quad\ \|f(y,t,\omega(x,t)) - f(y,t,\omega(y,t))\| \\ &\leq L_1 \cdot |x-y| + L_2 \cdot \|\omega(x,t) - \omega(y,t)\| \\ &\leq L_1 \cdot |x-y| + L_2 \cdot L_3 \cdot |x-y| = (L_1 + L_2 \cdot L_3) \cdot |x-y| \end{aligned}$$

and therefore, by Theorem 9.2,

$$\begin{aligned} \Phi(\omega)_t(x,t) - \Phi(\omega)_{xx}(x,t) &= f(x,t,\omega(x,t)), \quad (x,t) \in \mathbb{R} \times (0,T), \\ \Phi(\omega)(x,0) &= 0, \quad x \in \mathbb{R}. \end{aligned}$$

In particular, the function $\Phi(\omega)(x,t)$ is continuous on $\mathbb{R} \times [0,T]$. This means that (9.5) has a solution if Φ possesses a fixed point. In order to apply Tarski's fixed point theorem, we have to verify three things.

(A1) $(\Omega, \leq)$ is a complete lattice.

(A2) $\Phi : \Omega \to \Omega$; i.e., we have to show that Φ is a self-mapping of Ω.

(A3) $\omega \leq v \Rightarrow \Phi(\omega) \leq \Phi(v)$ for all ω, $v \in \Omega$.

Already in Example 9.2 we have shown the first point. Concerning (A2): We have to show $\Phi(\omega) \in \Omega$, if $\omega \in \Omega$; i.e., we have to show three things:

(B1) $-TM \leq \Phi_n(\omega)(x,t) \leq TM$ for all $n \in \mathbb{N}$ and for all $(x,t) \in \mathbb{R} \times [0,T]$.

(B2) $\|\Phi(\omega)(x,t) - \Phi(\omega)(y,t)\| \leq L_3 \cdot |x-y|$ for all $x, y \in \mathbb{R}$ and for all $t \in [0,T]$.

(B3) $\|\Phi(\omega)(x,t) - \Phi(\omega)(x,s)\| \leq L_4 \cdot |t-s|$ for all $x \in \mathbb{R}$ and for all $s, t \in [0,T]$.

Concerning (B1): Since $f(x,t)$ is bounded by M we get after substituting $\xi = x - \sqrt{4(t-\tau)}\,\alpha$ that

$$\begin{aligned} \|\Phi(\omega)(x,t)\| &\leq M \cdot \int_0^t \int_{-\infty}^{\infty} \tfrac{1}{\sqrt{4\pi(t-\tau)}} \cdot e^{-\dfrac{(x-\xi)^2}{4(t-\tau)}}\, d\xi\, d\tau \\ &= M \cdot \int_0^t \int_{-\infty}^{\infty} \tfrac{1}{\sqrt{\pi}} \cdot e^{-\alpha^2}\, d\alpha\, d\tau. \end{aligned}$$

Using $\int\limits_{-\infty}^{\infty} \frac{1}{\sqrt{\pi}} \cdot e^{-\alpha^2}\, d\alpha = 1$ we get

$$\|\Phi(\omega)(x,t)\| \leq M \cdot \int\limits_0^t 1 d\tau \leq T \cdot M.$$

Concerning (B2): For fixed $n \in \mathbb{N}$ we have

$$\Phi_n(\omega)_x(x,t) = \int\limits_0^t \int\limits_{-\infty}^{\infty} \frac{1}{\sqrt{4\pi(t-\tau)}} e^{-\frac{(x-\xi)^2}{4(t-\tau)}} \frac{\xi - x}{2(t-\tau)} f_n(\xi,\tau,\omega(\xi,\tau))\, d\xi\, d\tau$$

and we get $|\Phi_n(\omega)_x(x,t)| \leq \frac{M}{\sqrt{\pi}}(I_1 + I_2)$ with

$$\begin{aligned} I_1 &:= \int\limits_0^t \int\limits_{-\infty}^{x} \frac{1}{\sqrt{4(t-\tau)}} \cdot e^{-\frac{(x-\xi)^2}{4(t-\tau)}} \cdot \frac{|\xi - x|}{2(t-\tau)}\, d\xi\, d\tau \\ &= \int\limits_0^t \int\limits_{-\infty}^{x} \frac{1}{\sqrt{4(t-\tau)}} \cdot e^{-\frac{(x-\xi)^2}{4(t-\tau)}} \cdot \frac{x - \xi}{2(t-\tau)}\, d\xi\, d\tau \end{aligned}$$

and

$$\begin{aligned} I_2 &:= \int\limits_0^t \int\limits_{x}^{\infty} \frac{1}{\sqrt{4(t-\tau)}} \cdot e^{-\frac{(x-\xi)^2}{4(t-\tau)}} \cdot \frac{|\xi - x|}{2(t-\tau)}\, d\xi\, d\tau \\ &= \int\limits_0^t \int\limits_{x}^{\infty} \frac{1}{\sqrt{4(t-\tau)}} \cdot e^{-\frac{(x-\xi)^2}{4(t-\tau)}} \cdot \frac{\xi - x}{2(t-\tau)}\, d\xi\, d\tau. \end{aligned}$$

Substituting $\alpha := -\frac{(x-\xi)^2}{4(t-\tau)}$, $-\infty < \xi \leq x$ it follows (using $d\alpha = \frac{x-\xi}{2(t-\tau)}\, d\xi$) that

$$\begin{aligned} I_1 &= \int\limits_0^t \int\limits_{-\infty}^{0} \frac{1}{\sqrt{4(t-\tau)}} \cdot e^{\alpha}\, d\alpha\, d\tau = \int\limits_0^t \frac{1}{\sqrt{4(t-\tau)}} \cdot \left(\int\limits_{-\infty}^{0} e^{\alpha}\, d\alpha \right) d\tau \\ &= \int\limits_0^t \frac{1}{\sqrt{4(t-\tau)}}\, d\tau = [-\sqrt{t-\tau}]|_0^t = \sqrt{t} \leq \sqrt{T}. \end{aligned}$$

In the same way, substituting $\alpha := -\frac{(x-\xi)^2}{4(t-\tau)}$, $x \leq \xi < \infty$ it follows using the same change of variables that

$$
\begin{aligned}
I_2 &= \int_0^t \int_0^{-\infty} \frac{1}{\sqrt{4(t-\tau)}} \cdot e^{\alpha} \cdot (-1)\, d\alpha\, d\tau = \int_0^t \frac{1}{\sqrt{4(t-\tau)}} \cdot \left(\int_{-\infty}^0 e^{\alpha}\, d\alpha \right) d\tau \\
&= \int_0^t \frac{1}{\sqrt{4(t-\tau)}}\, d\tau = [-\sqrt{t-\tau}]|_0^t = \sqrt{t} \le \sqrt{T}.
\end{aligned}
$$

All together,

$$
|\Phi_n(\omega)_x(x,t)| \le \frac{M}{\sqrt{\pi}}(I_1 + I_2) \le \frac{M}{\sqrt{\pi}} \cdot 2 \cdot \sqrt{T} < 2 \cdot M \cdot \sqrt{T} = L_3.
$$

Using the mean-value theorem, there exists η between x and y satisfying

$$
|\Phi_n(\omega)(x,t) - \Phi_n(\omega)(y,t)| = |\Phi_n(\omega)_x(\eta,t) \cdot (x-y)| \le L_3 \cdot |x-y|.
$$

Since the right-hand side does not depend on n, we can conclude

$$
\|\Phi(\omega)(x,t) - \Phi(\omega)(y,t)\| \le L_3 \cdot |x-y|.
$$

Concerning (B3): Without loss of generality we assume $s < t$. Then, for $D := \|\Phi(\omega)(x,t) - \Phi(\omega)(x,s)\|$ we get

$$
\begin{aligned}
D \le & \left\| \underbrace{\int_0^s \int_{-\infty}^{\infty} \frac{1}{\sqrt{4\pi(t-\tau)}} \cdot e^{-\frac{(x-\xi)^2}{4(t-\tau)}} \cdot f(\xi,\tau,\omega(\xi,\tau))\, d\xi\, d\tau}_{=:J_{11}} \right. \\
& \left. - \underbrace{\int_0^s \int_{-\infty}^{\infty} \frac{1}{\sqrt{4\pi(s-\tau)}} \cdot e^{-\frac{(x-\xi)^2}{4(s-\tau)}} \cdot f(\xi,\tau,\omega(\xi,\tau))\, d\xi\, d\tau}_{=:J_{12}} \right\| \\
& + \left\| \underbrace{\int_s^t \int_{-\infty}^{\infty} \frac{1}{\sqrt{4\pi(t-\tau)}} \cdot e^{-\frac{(x-\xi)^2}{4(t-\tau)}} \cdot f(\xi,\tau,\omega(\xi,\tau))\, d\xi\, d\tau}_{=:J_2} \right\|.
\end{aligned}
$$

As in Case (B1) one can show $\|J_2\| \le M \cdot (t-s) = M \cdot |t-s|$. So, the hard work is kept in estimating J_{11} and J_{12}, respectively. Substituting $\alpha = \frac{x-\xi}{\sqrt{4(t-\tau)}}$

and $\alpha = \frac{x-\xi}{\sqrt{4(s-\tau)}}$, respectively, we get

$$\|J_{11} - J_{12}\| \leq \int_0^s \int_{-\infty}^{\infty} \frac{1}{\sqrt{\pi}} \cdot e^{-\alpha^2} \cdot \Big\| f(x - 2\alpha\sqrt{t-\tau}, \tau, \omega(x - 2\alpha\sqrt{t-\tau}, \tau))$$

$$- f(x - 2\alpha\sqrt{s-\tau}, \tau, \omega(x - 2\alpha\sqrt{s-\tau}, \tau)) \Big\| \, d\alpha \, d\tau$$

$$\leq \int_0^s \int_{-\infty}^{\infty} \frac{1}{\sqrt{\pi}} \cdot e^{-\alpha^2} \cdot (L_1 + L_2 \cdot L_3) \cdot \left| x - 2\alpha\sqrt{t-\tau} - x + 2\alpha\sqrt{s-\tau} \right| \, d\alpha \, d\tau$$

$$= \frac{1}{\sqrt{\pi}} \cdot (L_1 + L_2 \cdot L_3) \cdot \underbrace{\int_0^s (\sqrt{t-\tau} - \sqrt{s-\tau}) \, d\tau}_{=:F} \cdot \underbrace{\int_{-\infty}^{\infty} e^{-\alpha^2} 2|\alpha| \, d\alpha}_{=2} .$$

Concerning F we can conclude

$$F = \frac{2}{3}[-(t-\tau)^{\frac{3}{2}} + (s-\tau)^{\frac{3}{2}}]\Big|_0^s = \frac{2}{3}(t^{\frac{3}{2}} - s^{\frac{3}{2}} - (t-s)^{\frac{3}{2}}) < \frac{2}{3}(t^{\frac{3}{2}} - s^{\frac{3}{2}}) =: P.$$

Considering the function $b(x) := \frac{2}{3}x^{\frac{3}{2}}$, $0 \leq x \leq T$ with $b'(x) = \sqrt{x}$ by using the mean-value theorem we get

$$P = \frac{2}{3}(t^{\frac{3}{2}} - s^{\frac{3}{2}}) = b(t) - b(s) = b'(\eta) \cdot (t-s) \leq \sqrt{T} \cdot (t-s) = \sqrt{T} \cdot |t-s|$$

for some $\eta \in (s,t)$. All together it holds

$$\begin{aligned} D &\leq \|J_{11} - J_{12}\| + \|J_2\| \\ &\leq \tfrac{1}{\sqrt{\pi}} \cdot (L_1 + L_2 \cdot L_3) \cdot 2 \cdot \sqrt{T} \cdot |t-s| + M \cdot |t-s| \\ &= L_4 \cdot |t-s|. \end{aligned}$$

Concerning (A3): Let $\omega \leq v$. Since

$$\frac{1}{\sqrt{4\pi(t-\tau)}} \cdot e^{-\frac{(x-\xi)^2}{4(t-\tau)}} > 0$$

it follows by the assumption on f for all $n \in \mathbb{N}$

$$\Phi_n(\omega)(x,t) = \int_0^t \int_{-\infty}^{\infty} \frac{1}{\sqrt{4\pi(t-\tau)}} \cdot e^{-\frac{(x-\xi)^2}{4(t-\tau)}} \cdot f_n(\xi, \tau, \omega(\xi, \tau)) \, d\xi \, d\tau$$

$$\leq \int_0^t \int_{-\infty}^{\infty} \frac{1}{\sqrt{4\pi(t-\tau)}} \cdot e^{-\frac{(x-\xi)^2}{4(t-\tau)}} \cdot f_n(\xi, \tau, v(\xi, \tau))\, d\xi\, d\tau = \Phi_n(v)(x,t).$$

So, all assumptions of Tarski's fixed point theorem are fulfilled, and we can apply it. □

Finally, we want to note, that is not easy to see, how Theorem 9.3 can be proved using Schauder's fixed point theorem.

10 Further Reading

To emphasize that mathematical research never ends we add some more fixed point theorems for further reading. To avoid any kind of rate the following list is given in alphabetical order:

- The fixed point theorem of Browder, see (17).
- The fixed point theorem of Darbo, see (26).
- The fixed point theorem of Herzog and Kunstmann, see (45).
- The fixed point theorem of Kakutani, see (50).
- The fixed point theorem of Kneser, see (53).
- The fixed point theorem of Lagler and Volkmann, see (56).
- The fixed point theorem of Lefschetz and Hopf, see (30).
- The fixed point theorem of Lemmert, see (58).
- The fixed point theorem of Sadovskii, see (77).
- The fixed point theorem of Tychonoff, see (100).
- The fixed point theorem of Weissinger, see (108).

Some of them are (just) generalizations of others, some of them have also very nice applications. The reader might search for more fixed point theorems or the reader might extend the presented applications by his own or, even better, the reader will apply some fixed point theorem in a new way. Anyway, it's up to you now.

Bibliography

[1] R. A. ADAMS, *Sobolev spaces*, Academic Press, New York, 1975.

[2] M. AIGNER, G. M. ZIEGLER, *Proofs from THE BOOK*, Springer-Verlag, 2010.

[3] G. E. ALEFELD, X. CHEN, F. A. POTRA, *Numerical validation of solutions of complementarity problems: the nonlinear case*, Numer. Math., 92 (2002), pp. 1-16.

[4] G. ALEFELD, A. GIENGER, F. POTRA, *Efficient numerical validation of solutions of nonlinear systems*, SIAM J. Numer. Anal., 31 (1994), pp. 252-260.

[5] G. ALEFELD, G. MAYER, *The Cholesky method for interval data*, Linear Algebra Appl., 194 (1993), pp. 161-182.

[6] G. ALEFELD, P. VOLKMANN, *Regular splittings and monotone iteration functions*, Numer. Math., 46 (1985), pp. 213-228.

[7] G. ALEFELD, J. HERZBERGER, *Introduction to interval computations*, Academic Press, New York, 1983.

[8] G. ALEFELD, *Anwendungen des Fixpunktsatzes für pseudometrische Räume in der Intervallrechnung*, Numer. Math., 17 (1971), pp. 33-39.

[9] J. APPELL, M. VÄTH, *Elemente der Funktionalanalysis*, Vieweg, Wiesbaden, 2005.

[10] S. BANACH, *Sur les opérations dans les ensembles abstraits et leur application aux équations intégrales*, Fund. math., 3 (1922), pp. 133-181.

[11] W. BENZ, H. KARZEL, A. KREUZER, *Emanuel Sperner, Gesammelte Werke*, Heldermann Verlag, 2005.

[12] P. BOD, *On closed sets having a least element*, Optimization and operations research. Lecture Notes in Econom. Math. Systems, 117 (1976), pp. 23-34.

[13] P. Bod, *Minimality and complementarity properties of s.c. Z-functions and a forgotten theorem due to Georg Wintgen*, Math. Operationsforsch. Statist., 6 (1975), pp. 867-872.

[14] K. C. Border, *Fixed point theorems with applications to economics and game theory*, University Press, Cambridge, 1985.

[15] B. Breuer, P. J. McKenna, M. Plum, *Multiple solutions for a semilinear boundary value problem: a computational multiplicity proof*, J. Differential Equations, 195 (2003), pp. 243-269.

[16] L. E. J. Brouwer, *Über Abbildungen von Mannigfaltigkeiten*, Math. Annalen, 71 (1912), pp. 97-115.

[17] F. Browder, *On a generalization of the Schauder fixed point theorem*, Duke Math. J., 26 (1959), pp. 291-303.

[18] J. R. Cannon, *Multiphase parabolic free boundary value problems*, in: D. G. Wilson, A. D. Solomon, P. T. Boggs (eds.), *Moving boundary problems*, Academic Press 1978, pp. 3-24.

[19] R. Chandrasekaran, *A special case of the complementary pivot problem*, Opsearch, 7 (1970), pp. 263-268.

[20] Y. S. Choi, P. J. McKenna, *A mountain pass method for the numerical solutions of semilinear elliptic problems*, Nonlinear Anal. Theory Methods Appl., 20 (1993), pp. 417-437.

[21] L. Collatz, *Funktionalanalysis und Numerische Mathematik*, Springer-Verlag, Heidelberg, 1964.

[22] R. W. Cottle, J.-S. Pang, R. E. Stone, *The linear complementarity problem*, Academic Press, San Diego, 1992.

[23] J. Crank, *Free and moving boundary problems*, Clarendon Press 1984.

[24] J. Cronin, *Fixed points and topological degree in nonlinear analysis*, Amer. Math. Soc., Mathematical Surveys and Monographs, 1964.

[25] E. N. Dancer, S. S. Yan, *On the superlinear Lazer-McKenna conjecture*, J. Differential Equations, 210 (2005), pp. 317-351.

[26] G. Darbo, *Punti uniti in trasformazioni a codominio non compatto*, Rend. Sem. Mat. Univ. Padova, 24 (1955), pp. 84-92.

[27] K. Deimling, *Ordinary differential equations in Banach spaces*, Springer-Verlag Berlin 1977.

[28] K. Deimling, *Nichtlineare Gleichungen und Abbildungsgrade*, Springer-Verlag, Berlin Heidelberg 1974.

[29] J. Dieudonné, *Deux exemples singuliers d'équationes différentielles*, Acta Sci. Math., 12B (1950), pp. 38-40.

[30] J. Dugundji, A. Granas, *Fixed point theory*, Springer-Verlag, 2003.

[31] F. Facchinei, J.-S. Pang, *Finite-dimensional variational inequality and complementarity problems, Volume I + II*, Springer-Verlag, 2003.

[32] J. Franklin, *Methods of mathematical economics. Linear and nonlinear programming, fixed-point theorems*, SIAM, 2002.

[33] A. Friedman, *Partial differential equations of parabolic type*, Prentice-Hall, INC 1964.

[34] A. Frommer, F. Hoxha, B. Lang, *Proving the existence of zeros using the topological degree and interval arithmetic*, J. Comput. Appl. Math., 199 (2007), pp. 397-402.

[35] A. Frommer, B. Lang, *Existence tests for solutions of nonlinear equations using Borsuk's theorem*, SIAM J. Numer. Anal., 43 (2005), pp. 1348-1361.

[36] A. Frommer, B. Lang, M. Schnurr, *A comparison of the Moore and Miranda existence tests, Computing*, 72 (2004), pp. 349-354.

[37] A. Frommer, *Proving conjectures by use of interval arithmetic*, in: *Perspectives on enclosure methods*, Springer-Verlag, 2001, pp. 1-13.

[38] A. N. Godunov, *O teoreme Peano v banahovyh prostranstvah*, Funkcional'. Analiz Prilozén, 9 (1974), pp. 59-60.

[39] D. Hammer, *Eine Verallgemeinerung des Algorithmus von Chandrasekaran zur Lösung nichtlinearer Komplementaritätsprobleme*, Diplomarbeit, Universität Karlsruhe, 2006.

[40] R. Hammer, M. Hocks, U. Kulisch, D. Ratz, *Numerical toolbox for verified computing. Volume I: Basic numerical problems. Theory, algorithms, and Pascal-XSC programs*, Springer-Verlag, 1993.

[41] E. Hansen, G. W. Walster, *Global optimization using interval analysis. Second edition, revised and expanded. With a foreword by Ramon Moore*, Marcel Dekker, Inc., New York, 2004.

[42] E. R. Hansen, *Interval forms of Newton's method*, Computing, 20 (1978), pp. 153-163.

[43] E. R. Hansen, *On solving two-point boundary-value problems using interval arithmetic. In: Topics in interval analysis, E. Hansen (Editor)*, Clarendon Press, Oxford, (1969), pp. 74-90.

[44] P. T. Harker, J.-S. Pang, *Finite-dimensional variational inequality and nonlinear complementarity problems: A survey of theory, algorithms and applications*, Math. Programming, 48 (1990), pp. 161-220.

[45] G. Herzog, P. Kunstmann, *A fixed point theorem for decreasing functions*, Numer. Funct. Anal. Optim., 34 (2013), pp. 530-538.

[46] H. Heuser, *Lehrbuch der Analysis, Teil 2*, Teubner, Stuttgart, 1983.

[47] D. Idczak, M. Majewski, *A generalization of the Poincaré-Miranda theorem with an application to the controllability of nonlinear repetitive processes*, Asian Journal of Control, 12 (2010), pp. 168-176.

[48] I. James, *Remarkable Mathematicians, from Euler to von Neumann*, University Press, Cambridge, 2002.

[49] S. Kakutani, *Topological properties of the unit sphere in Hilbert space*, Proc. Imp. Acad. Tokyo, 19 (1943), pp. 269-271.

[50] S. Kakutani, *A generalization of Brouwer's fixed point theorem*, Duke Math. J., 8 (1941), pp. 457-459.

[51] J. B. Kioustelidis, *Algorithmic error estimation for approximate solutions of nonlinear systems of equations*, Computing, 19 (1978), pp. 313-320.

[52] R. Klatte, U. Kulisch, A. Wiethoff, C. Lawo, M. Rauch, *C-XSC. A C++ class library for extended scientific computing*, Springer-Verlag, 1993.

[53] H. Kneser, *Eine direkte Ableitung des Zornschen Lemmas aus dem Auswahlaxiom*, Math. Z., 53 (1950), pp. 110-113.

[54] R. Krawczyk, *Newton-Algorithmen zur Bestimmung von Nullstellen mit Fehlerschranken*, Computing, 4 (1969), pp 187-201.

[55] W. Kulpa, *The Poincaré-Miranda theorem*, Amer. Math. Monthly, 104 (1997), pp. 545-550.

[56] M. Lagler, P. Volkmann *Über Fixpunktsätze in geordneten Mengen*, Math. Nachr., 185 (1997), pp. 111-114.

[57] C. E. Lemke, J. T. Howson, *Equilibrium points of bimatrix games*, J. Soc. Indust. Appl. Math., 12 (1964), pp. 413-423.

[58] R. LEMMERT, *Existenzsätze für gewöhnliche Differentialgleichungen in geordneten Banachräumen*, Funkcial. Ekvac., Ser. Internac., 32 (1989), pp. 243-249.

[59] J. MILNOR, *John Nash and "A beautiful mind"*, Notices Amer. Soc., 45 (1998), pp. 1329-1332.

[60] C. MIRANDA, *Un osservatione su un theorema di Brouwer*, Belletino Unione Math. Ital. Ser. II, (1940), pp. 5-7.

[61] R. E. MOORE, J. B. KIOUSTELIDIS, *A simple test for accuracy of approximate solutions to nonlinear (or linear) systems*, SIAM J. Numer. Anal., 17 (1980), pp. 521-529.

[62] R. E. MOORE, *A test for existence of solutions to nonlinear systems*, SIAM J. Numer. Anal., 14 (1977), pp. 611-615.

[63] J. J. MORÉ, *Classes of functions and feasibility conditions in nonlinear complementarity problems*, Math. Programming, 6 (1974), pp. 327-338.

[64] S. A. MORRIS, *An elementary proof that the Hilbert cube is compact*, Amer. Math. Monthly, 91 (1984), pp. 563-564.

[65] M. T. NAKAO, *Solving nonlinear elliptic problems with result verification using an H^{-1} type residual iteration*, Computing (Suppl. 9) (1993), pp. 161-173.

[66] J. F. NASH, *Noncooperative games*, Ann. of Math., 54 (1951), pp. 286-295.

[67] M. NEHER, *The mean value form for complex analytic functions*, Computing, 67 (2001), pp. 255-268.

[68] J. M. ORTEGA, W. C. RHEINBOLDT, *Iterative solutions of nonlinear equations in several variables*, Academic Press, New York, 1970.

[69] G. OWEN, *Game theory*, W. B. Saunders Company, London, 1968.

[70] G. PEANO, *Démonstration de l'integrabilité des équations différentielles ordinaires*, Math. Ann., 37 (1890), pp. 182-228.

[71] M. PLUM, *Existence and multiplicity proofs for semilinear elliptic boundary value problems by computer assistance*, Jahresber. Deutsch. Math.-Verein., 110 (2008), pp. 19-54.

[72] M. PLUM, *Computer-assisted proofs for partial differential equations*, spring tutorial seminar on computer assisted proofs - numeric and symbolic approaches, 21st Century COE Program, Kyushu University, 2005.

[73] W. POUNDSTONE, *Prisoners' dilemma*, University Press, Oxford 1993.

[74] W. C. RHEINBOLDT, *On M-functions and their applications to nonlinear Gauss-Seidel iterations and to network flows*, J. Math. Anal. Appl., 32 (1970), pp. 274-307.

[75] S. ROBINSON, *The problem with blondes*, SIAM News, 35 (2002), p. 20.

[76] S. RUMP, *INTLAB - INTerval LABoratory. In: Csendes, Tibor (ed.), Developments in reliable computing*, (1999), pp. 77-104.

[77] B. N. SADOVSKII, *On a fixed point principle*, Funkcional. Anal. i Prilozen, 1 (1967), pp. 74-76.

[78] U. SCHÄFER, *A fixed point theorem based on Miranda*, Fixed Point Theory Appl., Vol. 2007, Article ID 78706 (2007), 5 pages.

[79] U. SCHÄFER, *On Tamir's algorithm for solving the nonlinear complementarity problem*, PAMM, Proc. Appl. Math. Mech., 7 (2007), pp. 2060057-2060058.

[80] U. SCHÄFER, *On computer-assisted proofs for solutions of linear complementarity problems*, J. Comput. Appl. Math., 199 (2007), pp. 257-262.

[81] U. SCHÄFER, *Wie erklärt man ein Nash-Gleichgewicht*, Elem. Math., 62 (2007), pp. 1-7.

[82] U. SCHÄFER, M. SCHNURR, *A comparison of simple tests for accuracy of approximate solutions to nonlinear systems with uncertain data*, J. Ind. Manag. Optim., Vol. 2, No. 4 (2006), pp. 425-434.

[83] U. SCHÄFER, *Unique solvability of an ordinary free boundary problem*, Rocky Mountain J. Math., 34 (2004), pp. 341-346.

[84] U. SCHÄFER, *Accelerated enclosure methods for ordinary free boundary problems*, Reliab. Comput., 9 (2003), pp. 391-403.

[85] U. SCHÄFER, *Aspects for a block version of the interval Cholesky algorithm*, J. Comput. Appl. Math., 152 (2003), pp. 481-491.

[86] U. SCHÄFER, *Two ways to extend the Cholesky decomposition to block matrices with interval entries*, Reliab. Comput., 8 (2002), pp. 1-20.

[87] U. SCHÄFER, *An existence theorem for a parabolic differential equation in* $l^\infty(A)$ *based on the Tarski fixed point theorem*, Demonstratio Math., 30 (1997), pp. 461-464.

[88] J. SCHAUDER, *Der Fixpunktsatz in Funktionalräumen*, Studia Math., 2 (1930), pp. 171-180.

[89] M. SCHNURR, *Computing slope enclosures by exploiting a unique point of inflection*, Appl. Math. Comput., 204 (2008), pp. 249-256.

[90] M. SCHNURR, D. RATZ, *Slope enclosures for functions given by two or more branches*, BIT, 48 (2008), pp. 783-797.

[91] M. SCHNURR, *On the proofs of some statements concerning the theorems of Kantorovich, Moore, and Miranda*, Reliab. Comput., 11 (2005), pp. 77-85.

[92] Y. A. SHASHKIN, *Fixed points*, American Mathematical Society, 1991.

[93] A. SIMON, I. OULD-AHMED-IZID-BIH, I. MOUTOUSSAMY, P. VOLKMANN, *Structures ordonnées et équations elliptiques semi-linéaires*, Rend. Circ. Mat. Palermo, 41 (1992), pp. 315-324.

[94] A. SIMON, P. VOLKMANN, *Équations elliptiques dans les espaces de Banach ordonnées*, C. R. Acad. Sci. Paris, t.315, Serie I, (1992), pp. 1245-1248.

[95] D. R. SMART, *Fixed point theorems*, University Press, Cambridge, 1974.

[96] E. SPERNER, *Neuer Beweis für die Invarianz der Dimensionszahl und des Gebietes*, Abh. Math. Sem. Hamburg, 6 (1928), pp. 265-272.

[97] A. TAMIR, *Minimality and complementarity properties associated with Z-functions and M-functions*, Math. Programming, 7 (1974), pp. 17-31.

[98] A. TARSKI, *A lattice-theoretical fixpoint theorem and its applications*, Pacific J. Math., 5 (1955), pp. 285-309.

[99] R. C. THOMPSON, *A note on monotonicity properties of a free boundary problem for an ordinary differential equation*, Rocky Mountain. J. Math., 12 (1982), pp. 735-739.

[100] A. TYCHONOFF, *Ein Fixpunktsatz*, Math. Ann., 111 (1935), pp. 767-776.

[101] R. S. VARGA, *Matrix iterative analysis*, Prentice-Hall, 1962.

[102] P. VOLKMANN, *Cinq cours sur les équations différentielles dans les espaces de Banach*, In: A. GRANAS, M. FRIGON (EDS.), *Topological methods in differential equations and inclusions*, Kluwer Dordrecht 1995, pp. 501-520.

[103] P. VOLKMANN, *Existenzsätze für gewöhnliche Differentialgleichungen in Banachräumen*, Tech. Univ. Berlin, Berlin 1985, pp. 271-287.

[104] M. N. Vrahatis, *A short proof and a generalization of Miranda's existence theorem*, Proc. Am. Math. Soc., 107 (1989), pp. 701-703.

[105] W. Walter, *Analysis II*, Springer-Verlag, 1990.

[106] W. Walter, *Gewöhnliche Differentialgleichungen*, Springer-Verlag, 1990.

[107] W. Walter, *There is an elementary proof of Peano's existence theorem*, Am. Math. Monthly, 78 (1971), pp. 170-173.

[108] J. Weissinger, *Zur Theorie und Anwendung des Iterationsverfahrens*, Math. Nachr., 8 (1952), pp. 193-212.

[109] F. Wilhelm, *Der Satz von Miranda und der Algorithmus von Tamir*, Diplomarbeit, Universität Karlsruhe, 2008.

[110] N. Yamamoto, M. T. Nakao, *Numerical verifications of solutions for elliptic equations in nonconvex polygonal domains*, Numer. Math., 65 (1993), pp. 503-521.

Index